THE LOCUM LIFE

A Physician's Guide to Locum Tenens

ANDREW N. WILNER, MD

ISBN: 978-1-4834-9466-1 (sc)
ISBN: 978-1-4834-9467-8 (hc)
ISBN: 978-1-4834-9465-4 (e)

Library of Congress Control Number: 2018914251

Cover Illustration: Anita White

Lulu Publishing Services rev. date: 12/20/2018

Dedication

Countless homemade muffins and cups of Barako,
served with endless patience, fueled the creation of
this book. Thank you, my beautiful wife, Irene.

Dr. Wilner provides the information that every doctor needs to know about locum tenens. Based on his personal hands-on experience as a locums physician, as well as his understanding of the growing need for locums doctors, he gives a realistic inside view of the pros and cons of locums work, an option that many doctors will consider at some point in their careers.

-Heidi Moawad, MD, Neurologist, Author,
Careers Beyond Clinical Medicine

As a medical services professional, I wish all those currently working or considering work as locum tenens would read this book. It would speed their applications through the hospital credentialing process.

-Kathy Matzka, CPMSM, CVPCS, FMSP

The Locum Life is an incredible book and has all the information I was looking for when I started my transition to full time locum tenens. Dr. Wilner provides excellent information in an easy to read manner. He addresses the advantages and disadvantages of The Locum Life. I will be recommending this book to anyone interested in starting The Locum Life.

-Jamie David Conklin, MD, Pulmonary/Critical Care,
Critical Breath MD, LLC

This thorough, accurate, balanced, and up to date introduction to locum tenens employment will prove insightful to clinicians at any stage in their career who are considering locums as a path to finding the perfect practice, achieving work/life balance, obtaining additional income, or transitioning to retirement.

-Eliza, MD, Dermatologist, Mother,
Financial Blogger at MinimalMD.com

Dr. Wilner's new book fills a much-needed gap in the career support literature for physicians by providing practical tips and advice to successfully motivate and navigate readers through the poorly understood world of locum life.

-Tiago Villaneuva, MD, Family Physician and Medical Editor

As a tax practitioner who works with physicians, including many who work locums, this book hits many of the financial and tax high points that are important when transitioning to self-employment. The information laid out mirrors many of the things I explain to new and prospective clients, so I think this will be a great resource to those exploring this income stream.

-Laura Clifford, CPA, President, Fox & Company CPAs, Inc

It's not a stretch to call The Locum Life a page-turner. As a professional career counselor and coach, I appreciated Dr. Wilner's encouraging, upbeat and realistic approach to this very unique medical career path. This is a highly-readable and practical guide that I can recommend to my clients, knowing they will gain a comprehensive understanding of the steps needed to thrive as locum tenens physicians.

-Amy Lindgren, President, Prototype Career Service

Andrew has written an infinitely readable, accessible and useful guide for any physician contemplating work as a locums physician. He identifies all of the significant issues involved and discusses the pros and cons of this type of employment. I wish I had a guide like this when I performed locums work earlier in my career.

-Steven Dale Boggs, MD, FASA, MBA, Professor,
Department of Anesthesiology

In the age of alternative careers, medical schools have not talked much about the "alternate" career of the physician locum tenens. This book fills this gap and provides a roadmap for those physicians interested in exploring alternative work arrangements that include health care delivery.

-Jose E. Cavazos, MD, PhD,
South Texas Medical Scientist Training Program

ALSO BY THE AUTHOR

Bullets and Brains (2013)

Epilepsy in Clinical Practice (2000)

Epilepsy: 199 Answers (1996)

ACKNOWLEDGMENTS

Writing is lonely work, and one brain can only accomplish so much. I am grateful to those fellow physicians, recruiting agents, and administrators who graciously shared their experiences and insights. Through many drafts and revisions, our discussions propelled my research and writing. More importantly, these diverse viewpoints widened the book's scope from a mere memoir of one physician's experiences to a comprehensive resource for physicians considering locum tenens.

I would like to thank the following people who shared their comments, critiques and expertise to help make *The Locum Life: A Physician's Guide to Locum Tenens*, a better book:

Nickolas D'Agostino, Gordon Banks, MD, Steven Boggs, MD, Lynette Bui, MD, Laura Bruse, MD, Jose Cavazos, MD, PhD, Laura Clifford CPA, Jamie Conklin, MD, Ken Donovan, MD, Eliza, the MinimalMD, Cory S. Fawcett, MD, Heather Feuerbacher, Haily Fowler, Amanda Gerber, MD, Marti Haykin, MD, Val Jones, MD, Kathy King, RDN, Janene Kingsley, Mack Land, MD, Michael Lewitt, MD, Amy Lindgren, Mike Lucas, Kathy Matzka, John McBurney, MD, Heidi Moawad, MD, Tony Mosley, MD, Alexi Nazem, MD, Jon Olson, MD, Aurora Pajeau, MD, David Sarmiento, MD, Angie Scarle, Rob Scott, MD, Karen Shackelford, MD, Louis Spikol, MD, Calvin Spott, MD, Stacey Stanley, Bret Stetka, MD, Spencer Sutherland, Amy Wecker, MD, Eric Wilner, MD, Irene P. Wilner, Colin Zhu, DO.

DISCLAIMER

Introduction

The Locum Life: A Physician's Guide to Locum Tenens is a compilation of information to inform and advise physicians considering locum tenens work. While the publisher and I have employed our best efforts, the book's contents do not intend to provide professional business, medical, or legal advice and should not replace the counsel of an accountant, physician, or attorney, respectively. Neither the publisher nor I assume liability for loss or damages resulting from its use.

I have enjoyed and profited from my locum tenens assignments. However, my past experience is no guarantee of your future results. Locum tenens opportunities are as diverse as the physicians who perform them. Your experiences may differ significantly from mine.

No Endorsements

From time to time, brand or company names may be mentioned. For example, I refer to Dropbox and discuss it along with iCloud Drive, Microsoft OneDrive and other services. Veridoc, a service that provides primary source verification for state licenses, is mentioned as part of a long chapter on credentialing. I discuss these products because I've used them, and they work for me. However, there may be better alternatives that I haven't found yet. It is not my intention to endorse any specific agency, company or product.

Truth or Fiction

All of my experiences related in the following chapters, as unlikely as some may seem, really happened. If the revelation of locations or identities of those involved would cause undue embarrassment, these details have been omitted.

Current Information

While every effort has been made to ensure that the information provided is accurate and timely, it comes without guarantees. In the fast-moving world of U.S. health care, some information may be outdated by the time you read your copy.

The information in this book serves as an introduction to the nuts and bolts of leading a successful locum tenens career. Basic material broaches complex topics such as professional liability and taxes. This information should prepare you to better utilize your attorney, certified public accountant, and other professional advisors.

Conflicts of Interest

The writing of this book has not been subsidized by any angel investor, health care facility, Kickstarter project, or locum tenens agency. I have no relevant conflicts of interest to declare. While I appreciate the kind assistance of those mentioned in the acknowledgments, all errors and omissions are my own.

INTRODUCTION

Thanks for picking up this book. If you are considering working locum tenens, *The Locum Life: A Physician's Guide to Locum Tenens*, will get you started.

Please allow me to introduce myself. I'm a board-certified neurologist, board-certified internist, fellowship trained epileptologist, medical journalist and practicing physician. Locum tenens has proved the best solution to balance my two lifelong careers of medicine and writing, provide time for family and hobbies, and pay the bills. Over the years, the temporary nature of locum tenens work provided me the freedom to write several books, author hundreds of articles, lead medical missions in the jungles of the Philippines, and film fascinating sea creatures underwater.

I took my first locum tenens assignment in 1982 as a part-time emergency room physician. In between suturing lacerations, delivering babies, extracting foreign bodies and treating motor vehicle accident victims, I wrote my first manuscript on an Olivetti electric typewriter. I also discovered that patients with neurologic symptoms fascinated me and set me on course to become a neurologist and epilepsy specialist.

During more than 35 years as a licensed physician, I've practiced in a variety of medical environments including inpatient, outpatient, academic and private practices. In order to obtain these diverse positions, I've worked with locum tenens agencies and independently. In the process, I've filled out countless forms, submitted to FBI criminal background checks, garnered 13 state licenses and mastered vital secretarial skills.

While numerous articles on the internet about locum tenens exist, full-fledged, authoritative books are rare. *The Locum Life: A Physician's Guide to Locum Tenens*, is the only one written by a neurologist with extensive locum tenens and traditional medical practice experience who also happens to be a professional writer. *The Locum Life* is the book I wish I had before starting my first locum tenens job nearly four decades ago.

Physicians thinking about trying a locum tenens assignment need to be aware of potential benefits and pitfalls. *The Locum Life* defines locum tenens, discusses motivations for locums work, its advantages and disadvantages at different stages of one's career, the pros and cons of staffing agencies, business of medicine, credentialing for hospital privileges and state licenses, professional liability insurance, as well as information regarding the mobile office, continuing medical education, travel tips and strategies for success. Pragmatic advice is plentiful, such as how to obtain the best assignments, streamline business tasks, and avoid malpractice suits. Chapters conclude with one or more practical recommendations.

It is not necessary to read the book cover to cover; each chapter stands alone. Start with the most relevant sections for your situation. As a result of this construction, the careful reader may observe some overlap of material amongst chapters.

An appendix includes a glossary of acronyms and a list of locum tenens agencies. As this is not a textbook, references appear sparingly. They are organized by chapter and included at the end of the book.

My favorite chapter is the one I didn't write! "Tales from the Trenches" includes frank, first-hand reports from other locum tenens physicians obtained through in-depth telephone and email interviews. These physician voices complement my own and make for a fascinating visceral read. If you're not sure whether locum tenens is for you, begin your research here. If you're still interested, the rest of the book will serve as a comprehensive guide for your new adventure.

Best wishes for a successful locum tenens career!

FOREWORD

The history of modern locum tenens goes back about 40 years in the United States. I've been part of it for almost 30.

During that time, I've seen just about every part of the industry, from recruiting physicians to running two of the largest locum tenens companies, CompHealth and Weatherby Healthcare. I was also one of the first presidents of the National Association of Locum Tenens Organizations (NALTO), the association responsible for setting industry standards and ethical guidelines.

I've watched locum tenens grow into a multi-billion dollar industry that places physicians of every specialty in facilities across the country. But it's not just the industry that has changed. How doctors use temporary assignments has also evolved.

What was once mostly for older physicians transitioning from full-time work into retirement, has now become a way of life at all career phases. More and more, young physicians are using locum tenens right out of residency to figure out where and how they want to work. Mid-career physicians are working locum tenens to earn extra money, combat burnout, or step away from the "business" of medicine. Other physicians have even decided to turn these temporary jobs into full-time careers.

However, there are a few things that haven't really changed. Though more physicians are working locum tenens than ever before, the vast majority of doctors still only have a basic understanding of the concept.

With *The Locum Life: A Physician's Guide to Locum Tenens*, Dr. Wilner gives an unbiased look at locum tenens-how it works, why it matters, what it can do for physicians and facilities, and even why it's not for everyone.

A doctor once told me there's an unspoken physician code-doctors don't lie to other doctors. *The Locum Life* is the inside scoop, from one doctor to another.

In this book, Dr. Wilner answers just about every question you'd ever have about locum tenens. If you still have more, there's probably only one thing left for you to do: Try it for yourself.

David Baldridge
Chief Strategy Officer, CHG Healthcare

CONTENTS

CHAPTER 1
What is Locum Tenens?

Locum Tenens

One filling an office for a time or temporarily taking the place of another—used especially of a doctor or clergyman

Dream Job

Imagine a dream job where you could work when you want, where you want, how you want, ignore local politics, enjoy generous compensation, and bask in appreciation from patients, peers, and even administrators. That's a quick summary of *The Locum Life*!

Locum tenens physicians serve as "placeholders," taking the place of physicians on maternity, paternity or sick leave, away at medical conferences or vacation. Expanding medical clinics or hospitals also hire locum tenens physicians to fill in while they recruit for permanent staff.

Physicians may work locum tenens as a one-time experience, at different stages of their career, or as a full-time employment strategy. Assignments can be domestic or international and last as little as a day or longer than a year.

Locum tenens physicians allocate time to their families and hobbies as they see fit rather than squeeze their lives into the rigid schedule of a conventional job. It's all about flexibility.

An Expanding Industry

If you have been thinking about temporary work, you are not alone. In a recent survey, 11.5% of physicians indicated they would work locum tenens in the next 1-3 years, an increase from 9.1% in 2014 (Physicians Foundation 2016). To address physician interest, articles about locum tenens have appeared in *Family Practice Management, KevinMD.com, MD Magazine, Medpagetoday.com, Medscape.com, Medical Economics,* and the *New England Journal of Medicine* Career Center. Locum tenens radio programs regularly play on ReachMD.com. There are even locum tenens blogs. An article on locum tenens published in Medscape.com garnered more than 4,000 views (Wilner 2016).

Locum tenens is a growing multibillion-dollar industry. The 2017 Survey of Temporary Physician Staffing Trends revealed that demand for locum tenens physicians has never been greater (Staffcare 2017). In 2016, approximately 50,000 U.S. physicians worked locum tenens.

Three-quarters of all hospitals rely on locum tenens physicians to stay fully staffed. Needs are greatest for anesthesiology, family practice, general surgery, internal medicine, and psychiatry, but locum tenens opportunities exist for every specialty.

Physician as Free Agent

The growing physician shortage, estimated at 121,000 by 2030 by the Association of American Medical Colleges (AAMC), is an important factor fueling locum tenens growth. Another driver is the fact that more than 50% of physicians now work as employees for medical groups or hospitals. These physicians have far more professional mobility than traditional private practice owners who suffer constraints of office mortgages and monthly overhead. Since modern practices grow primarily via inclusion in insurance networks, physicians no longer have to invest as much energy to garner referrals. Consequently, physicians are no longer bound as tightly to their communities and can work as "free agents" in the national and even global marketplace. Like a freelance journalist, a locum tenens physician plies his or her trade for different employers.

Why Me?

Locum tenens may not be for you. Many physicians receive adequate professional and financial satisfaction from conventional full-time employment and have no desire to supplement or replace it. Besides, how many doctors dreamed of becoming a "placeholder" when they grew up?

However, for physicians looking for additional income or those who want to try out different practice styles or geographic locations, locum tenens merits a closer look. About 25% of locum tenens physicians hold permanent positions but boost their income with occasional locum assignments. In a recent survey, locum tenens physicians ranked "freedom and flexibility" as the number one advantage of locums, followed by "pay rate," "no politics," and "travel" (Staffcare 2017).

One locum tenens physician wrote, "The 4 Lessons of Locums. 1) Locums makes you MONEY...FAST! 2) Locums lets you explore other jobs and network to find other positions, 3) As Locums, you are paid for what you work, and 4) Locums let YOU decide what working conditions you're willing to accept and for what price" (Barron 2013). Sound good?

Too Young or Too Old?

Think you're too young? Some physicians choose locums as their first job out of residency. Too old? Not likely. Three-quarters of locum tenens physicians (including the author), reside on the far side of 50. Locum tenens is appropriate for young physicians who have just completed training (Chapter 5), mid-career physicians seeking new experiences (Chapter 6), and late-career physicians searching for part-time work in lieu of retirement (Chapter 7). Dual physician couples and others with demanding schedules may benefit from part-time locums. Locum tenens provides an attractive alternative for physicians at all career stages who value increased flexibility and improved work/life balance.

Travel

A long locum tenens tradition speaks to balancing travel and work. Alan Kronhaus, MD, the founder of Kron Medical, one of the first locum tenens companies, worked six months a year in Nevada. He spent the remaining six months enjoying the Colorado ski slopes, unfettered by clinical responsibilities.

Locum tenens physicians can literally travel the world while earning a paycheck. Although frequent travel may be more suitable to young, single physicians, locums assignments may also be attractive to men and women with children. International opportunities for U.S. physicians extend to the U.S. Virgin Islands as well as other locations such as Australia, Bermuda, Canada, Caribbean, China, Malaysia, New Zealand and others. Compensation in foreign countries might be lower than U.S. rates, but the travel and cultural experiences that families share come free!

An Army of One

All this freedom comes at a price. The independence and flexibility of locum tenens work imposes its own set of responsibilities including learning the fundamentals of running a business. For example, usually locum tenens physicians work as self-employed "independent contractors." Income is tallied on IRS 1099-MISC forms. Alternatively, locum tenens physicians may be traditional employees, with income reported on a W-2 form. I've had it both ways.

It's up to you to understand the difference between an IRS 1099-MISC and W-2 and manage your finances accordingly (Chapter 14). As a self-employed individual, you will be responsible for paying quarterly tax estimates, obtaining health insurance, contributing to a retirement plan, tallying deductible business expenses, and most importantly, creating your own work schedule. Many locum tenens physicians gladly accept these new responsibilities in return for the flexibility and control they gain over their professional and personal lives (Chapter 20).

Recommendations

Locum tenens is not for everyone. However, approximately 50,000 physicians have discovered that a flexible schedule enhances their lives. If you are curious about locum tenens and not completely satisfied with your current employment, read the rest of this book. By the time you finish, you should have a good idea whether locum tenens is for you.

CHAPTER 2

A Brief History of Locum Tenens

It Starts with Hippocrates...

Although the formal profession of locum tenens medicine is relatively new, it's safe to say that locum tenens physicians have been around for a long time. Records are scanty, but we know that Hippocrates taught the Art of Medicine on the Greek island of Kos. He also traveled extensively and provided temporary services. In terms of work/life balance, it looks like Hippocrates got it right more than 2,000 years ago.

During the 1800s, traveling physicians patched up pioneers settling the Wild West. Civil War military physicians treated combatants but also cared for people in local communities. In 1911, government funding encouraged physicians to travel to Native American tribal areas (Locumstory.com).

Modern Day

The creation of Médecins Sans Frontières (Doctors Without Borders) in 1971 formalized the concept of temporary volunteer doctors. About the same time, Therus Kolff and Alan Kronhaus initiated a program that recruited temporary physicians to underserved rural areas in the Western U.S. (Chapter 1). These efforts led to the formation of CompHealth, the first national locum tenens agency.

A Growth Industry

By 2002, more than 25,000 doctors worked locum tenens assignments. This number has swelled to approximately 50,000 today. In the U.S., locum tenens doctors see 20,000,000 patients a year, making a significant contribution to health care delivery. A recent survey revealed that the growing physician shortage coupled with physician willingness to consider locum tenens predicts continued growth of the locum tenens industry (Staffcare 2017). More than 100 locum tenens agencies recruit physician "placeholders" for temporary positions (Appendix 2).

We don't know for sure, but Hippocrates may have relied on local locum tenens physicians to cover his medical practice at the Asclepeion when he went on vacation.

Recommendations

The modern practice of locum tenens evolved from a long history of adventurous physicians helping patients in far-flung regions. Consider whether you're cut from the same cloth.

CHAPTER 3

Advantages of Locum Tenens

You're the Boss

One of the beauties of a locum tenens practice is the ability to create your own schedule. You decide when, where, how often, and how hard to work. It's one of the most flexible jobs in the world. Let's review the advantages:

Contracts?

All you have to do is show up and work. Contracts with locum tenens companies are simple; a few pages of boilerplate followed by an assignment letter that spells out duties and pay rate. No restrictive covenants. No lawyer to scrutinize the fine print. Sign and go to work.

Office Administration?

No billing. No payroll. No rent. No staff. No worries. Just check your bank account now and then to track direct deposit paychecks.

When?

You are master of your universe. How you balance your income, expenses, and free time is completely up to you. You can work one day a year or all 365. You can work weekends, every other week, or six months a year like Dr. Kronhaus. The possibilities are endless. Except perhaps for your immediate family, no one cares.

For some physicians, locums is a golden opportunity to work hard and often in order to reap impressive financial rewards. For me, the freedom to interrupt clinical work for several months at a time made locum tenens a pragmatic option. This unparalleled autonomy allowed travel to medical conferences to work as a medical journalist, lead annual medical missions, shoot underwater videos in Southeast Asia, and write my last book, *Bullets and Brains*. It's been a great ride!

Where?

Locum tenens assignments exist across the nation and internationally. As you might imagine, opportunities in cosmopolitan cities like Miami, New York, and San Francisco appear less often than in rural regions with less name recognition. For some assignments, I've had to consult a map to locate a hospital off the beaten path.

For example, I've worked in the dead of winter in what I consider a "remote region," Sioux Falls, South Dakota. I didn't know the thermometer could get that low...

But not all locums opportunities take you to out of the way locations. I've had assignments in the Boston suburbs, coastal Connecticut, downtown Minneapolis and sunny southern California.

Because each state requires its own medical license and each hospital performs its own credentialing, these prerequisites must be obtained prior to each assignment. Staffing agencies assist and reimburse license and credentialing fees.

Housing

In Sioux Falls, I stayed in a modest hotel off the highway and drove my rental car about 20 minutes to the hospital each day. In Minneapolis, I walked three blocks to work from a friendly, but somewhat shabby downtown hotel. In both cases, the staffing agency selected the hotel and paid for it. I arranged my New London position without an agency, so the rent for my cozy waterfront apartment a few blocks from the hospital came directly from my pocket. In southern California, a 30 minute drive from the beach brought me to a cute little hospital.

If you wish to see different parts of the country, maybe explore a city for a possible permanent practice, or combine work and vacation, you can try locums and get paid for your travel.

Travel Perks

My usual work routine is 7 days on/7 days off. This often entails at least one flight a week and prolonged hotel stays. All travel expenses are paid by the agency, but frequent flier miles as well as hotel and rental car award points are mine to keep. In a recent cross-country pleasure trip, I made good use of my Best Western "Diamond" status and thousands of reward points from long stays in Minneapolis. I'm hoarding frequent flier points now for another dive trip to the Philippines.

Compensation?

Compensation typically includes an hourly wage, housing, on call coverage, overtime, rental car, and round-trip transportation to the work site (Chapter 13). Earnings tend to equal or exceed permanent salaries because the hiring organization, usually a hospital or clinic, has an immediate and pressing need. The pay also needs to be higher because it doesn't include benefits like health insurance or pension contributions. The hourly rate is usually negotiable, and you have the final say. It's wonderful to say "yes" when you like the terms, and it's just as satisfying to say "no" when you don't. In today's market, there are always other locums opportunities.

It is possible to earn just as much working locum tenens if not more than in a traditional permanent position. Such was the case in one of my recent assignments. Because of its hourly pay structure and umpteen hours, I earned more in one year than ever in my life. I'm sure I made someone at the internal revenue service happy. I was truly sorry to see that position end!

Interestingly, had I been a full-time, permanent employee at the same facility, doing the same job, I would've earned less than I did as a locums. As a desperately needed, temporary employee, I possessed special value. Such is the strange world of medical economics.

Retirement

As a self-employed physician, you wear two hats; employer and employee. Your role as employer offers attractive retirement options unavailable to salaried employees. You may be eligible to start your own 401k retirement plan and make sizable contributions—as much as 25% of your income. Do yourself a favor and consult a certified public accountant (CPA) who specializes in physician clients (Chapter 14). Many physicians benefit from the counsel of a financial planner as well.

Financial Education

I have never had much interest in the financial world. Except for one college accounting course, which I barely passed, I have no formal business training. Balance sheets, income statements, shareholder equity statements, and the arcane rules of the U.S. tax code governing 401a's, 401k's, 403b's, and 457 plans hold no magic for my soul. Recently, 529s came to my attention, but that's another story!

Nonetheless, my personal portfolio now includes most of the above three-number retirement vehicles, and I am duty bound to understand them. Luckily, our online, interconnected world provides helpful self-education opportunities. For example, financial institutions like Fidelity, Schwab and Vanguard offer video tutorials. If you open an account, someone on the other end of an 800 line will patiently answer your questions.

My business education has also benefited from financial websites. In particular, I've discovered four helpful blogs, written by physicians for physicians; "Passive Income MD," "Physician on FIRE," "The White Coat Investor," and the "Wall Street Physician." These authors address common questions regarding investing, saving, spending and taxes. One even responded to a question I posted.

Many other online resources as well as self-help books can assist physicians faced with the challenges of self-employment. Of course, no matter how much I learn on my own, I always run important decisions past my CPA.

While the rules of finance may be confusing, it's not rocket science. If you can master the Krebs cycle and the intricacies of the hypothalamic-pituitary-adrenal axis, running your own business should be child's play.

Appreciation from Patients and Peers

Locum tenens positions exist because of an urgent clinical need that can't be satisfied any other way. Examples include a clinic with patients scheduled several months out, a chronically short-staffed ER with unacceptable wait times, or a recently expanded clinic paying rent on idle exam rooms. Unexpected physician departures are not uncommon and can rapidly fling a marginally managed practice into chaos.

In my experience, patients appreciate a locum tenens physician. For example, my presence at a short-staffed tertiary care center in Arizona decreased the clinic wait from 6 months to 2 weeks. Patients were overjoyed when they learned a new doctor had arrived. They thought I was great just by showing up!

The permanent staff physicians also welcomed me with open arms. Naturally, I attributed this warm reception to my stellar reputation and exemplary diagnostic skills. I learned later that three neurologists had abruptly abandoned the institution due to a contract dispute. Their busy administrative, clinical, and on call responsibilities were unceremoniously dumped on the hapless neurologists left behind. Given the circumstances, any neurologist who could share the load was welcome indeed!

I have never experienced "negative vibes" because of my locum tenens status. Let's face it, your peers are mostly concerned with getting their own work done. In locum situations, your colleagues are generally overworked, forced to take on the burden of an expanding practice or the workload of a missing colleague who has left temporarily or permanently. If you make their jobs easier, you will be valued. Further, because you're just passing through, you don't threaten the pecking order.

Administrators resent paying a premium for locums, but since it's a temporary situation, they can justify the additional expense. More importantly, a doctor in place addresses one of a manager's greatest fears: losing patients to the competition. Hospital and clinic administrators have always treated me with civility and even a bit of coddling. I suspect this beneficence is largely self-serving. If I was unhappy and left, they would have to start their search for a new "placeholder" all over again. From a bean counter's point of view, the imperfect solution of a transient, overpaid physician is preferable to no physician at all!

Local Politics

Mastering internecine politics may be a necessary evil when it comes to succeeding in a hospital environment. But as a locum tenens physician, you're above the fray. When coworkers gripe about the unfair call schedule, overbooked clinics, or broken coffee machine, you can listen and nod sympathetically. Then walk away and smile. It's not your problem. Your assignment has an endpoint, and you will shortly escape local injustices. All you have to do each day is complete your work, stay out of trouble, and collect your paycheck.

Recommendations

Every life choice offers advantages and disadvantages. In my career, the advantages of locum tenens have usually outweighed the disadvantages. The flexibility of locum tenens employment is unparalleled. However, I am not blind to the disadvantages, which are addressed in the next chapter. Learn about both before you take your first assignment.

CHAPTER 4

Disadvantages of Locum Tenens

Look Before You Leap

The Locum Life offers spectacular advantages over conventional full-time physician employment (Chapter 3). However, before you jump in, here's the flip side of the coin.

When?

It is completely up to you when to work. That's the good news. But once you've decided when you want to work, you must find an assignment that fits your time frame. While on assignment, you need to start thinking and planning for the next one.

How?

You will probably sign on with at least one staffing agency. I regularly use two of the largest, CompHealth and Staffcare. There are many others (Appendix 2). Agents work primarily on commission, so they're happy to serve you.

As each agency maintains a stable of doctors, plum assignments go fast, even the same day they become available. You must be prepared to respond quickly when an opening appeals to you, or better yet, encourage your agent to call with attractive positions before they're posted. A close working relationship with a staffing agent leads to the best assignments.

Paperwork Perdition

Getting started with an agency requires paperwork up front (Chapter 8). In addition, each new assignment requires appropriate state licensing and hospital privileges (Chapter 9). Obtaining both of these demands extraordinary diligence, patience, and persistence. Frustrated doctors have been known to abandon half-finished applications, preferring to forfeit an assignment rather than complete onerous and seemingly endless paperwork. I have been close to calling it quits on an application more than once.

You also need to obtain professional liability insurance, either on your own or through an agency (Chapters 11 and 12). This will usually require an additional application. If you arrange your own insurance, the cost is yours.

There's also the matter of presenting your credentials to the relevant insurance networks to obtain payment for your clinical services. Administrators will take care of most of this, but expect emails requesting yet more information as well as updates for a centralized website such as the Council for Affordable Quality Healthcare (CAQH).

To accomplish these tasks, your secretarial skills will need to be top notch. Because of the detailed and personal nature of these applications, it is difficult to outsource these chores to an administrative assistant. Agencies will assist as best they can.

Licensing

Some states make it easy. For example, North Dakota, a popular state for locum tenens, may issue a temporary license in as little as two weeks.

Permanent licenses always take longer. Three months would be quick, and six months or more is not unusual. For example, my recent Tennessee license application was filed in May, and I was scheduled to begin work in September. The license wasn't issued until November! I had to postpone my move from Arizona and survive on savings as the process dragged on.

For many states, initial application paperwork necessitates a week or more of uncompensated time. That's 40 hours at your desk filling in detailed biographic information already on your resume, which must be reformatted to each state's byzantine, proprietary application.

You may require clarification to comprehend cryptic application questions. Make sure you have a comfortable chair and a cup of coffee when you call the license board. Lengthy hold times are the norm. Be prepared to get disconnected while waiting and start all over again.

Administrative tasks abound. Some forms have to be notarized, and you may need to submit affidavits. Expect to contact the National Practitioner Data Bank to request a report.

You must submit fingerprints for a criminal background check. This requires one or more trips to the local police station, always a delight (Chapter 10).

An in-person interview may be required for state licensing. I spent three uncompensated days traveling to Mississippi for an interview and a medical society sponsored morality lecture. (The lecture was for everyone, not just me!)

Get your checkbook out. There are fees for many of these requirements. Track every dollar on a spreadsheet (Chapter 15). If you work with an agency, they will eventually reimburse you for license expenses.

DEA

You will need a Drug Enforcement Agency (DEA) registration number to prescribe controlled substances (Chapter 9). If assignments take you to different states, you must transfer your DEA registration from one state to the other. If you work in more than one state concurrently, you will need more than one DEA registration number. At one time, while working off and on in several states, I held three DEA numbers at $731 each (Chapter 9).

As a superfluous layer of bureaucracy, some, but not all, states require a controlled substance registration in addition to a federal DEA registration. These go by two acronyms; CSP (Controlled Substance for Practitioners) or CSA (Controlled Substance Act). You will have to check with your state to see whether you need one. If so, be prepared for a fee. My CSA for South Dakota cost $150.

In an effort to combat the opioid epidemic, many states have instituted a mandatory prescription monitoring program (PMP). Also known as a prescription monitoring and reporting system (PMRS), this database logs patients' controlled substance prescriptions. Physicians are required to register and look up each patient's record before prescribing a controlled substance. I haven't encountered a registration fee for this program, but signing up is one more thing you'll have to do each time you practice in a new state. PMPs are state specific, so prescriptions written in other states don't show up. Needless to say, a single federal system would be far more efficient and effective, but these two adjectives rarely apply to government programs.

Controlled substance licensing and regulations intend to protect patients from overprescribing and prevent diversion of substances with a high potential for abuse. However, they result in an impressive administrative burden for the locum tenens practitioner who works in more than one state.

Hospital Privileges

Once state licensing is complete, it's time to apply for hospital privileges. It would save time if state license and hospital privilege applications could be reviewed concurrently, but hospital credentialing committees won't even glance at your application until you are licensed in their state.

Assuming no black marks on your resume, obtaining hospital privileges should be pro forma. However, like state licensing, this process is guaranteed to be unduly time consuming and frustrating (Chapter 9). Many hospital boards meet only once a month, so two or three months of waiting is not unusual while hospital clerks painstakingly duplicate the work of the state license board, reconfirming all the fascinating factual information in your application.

If you've just graduated from residency, these applications will be less formidable. If you've had many locums assignments, current requirements to verify "everything" inevitably lead to prolonged delays.

Timing

While you may be ready to start work tomorrow, anticipate a wait time of at least 3-6 months before you actually obtain a state license and hospital privileges. Unfortunately, by the time you are properly licensed and credentialed, your enticing locum tenens opportunity may no longer exist.

Locum tenens positions come and go, that's the nature of the beast. In my own career, bureaucratic licensing and credentialing delays resulted in postponements of two assignments and cancellation of another. These missed opportunities translated into thousands of dollars of lost income. The process is wasteful beyond belief.

One might surmise that a hospital, in dire need of an additional physician, would expedite its own privileging process. Logical as this may seem, it does not happen. For administrators, the value of the process far exceeds the importance of the result. My experience suggests that hospital administrators place much more weight on preserving their Joint Commission seal of approval than on addressing immediate patient care needs.

Administrators follow the letter of the law even if it means excessive hiring delays or not hiring you at all because of a technicality. The latter situation happened to me once and it wasn't pretty. The details could fill a book...a story for another time!

Staffing agents are well aware of delays inherent in licensing and hospital privileging. A good agent will encourage you to complete all necessary paperwork as quickly as possible. They will assist with secretarial work as well.

Good agents will discourage you from taking assignments that require a new state license if they don't think it will arrive soon enough. A "bad" agent, desperate for a commission, will push you to take any assignment, whether it makes sense for you or not. When I first started out in locum tenens and didn't know any better, I suffered the latter experience. Needless to say, I wised up and changed agents.

Cancellations

Standard locum tenens contracts allow for cancellations up to 30 days before the start of an assignment by either party, *for any reason*. Hospitals and clinics would much rather have permanent employees. If they can find one sooner than expected, they don't want to be stuck with an expensive locums doc like you.

Last-minute cancellations are a part of life for locum tenens physicians and can wreak havoc on financial and vacation planning (Chapter 20). Other locums opportunities may or may not be available in that time-frame to fill the gap, and you may have to scramble to find one. Contracts promise compensation if a cancellation occurs within 30 days of the start date, but this covenant may not always be honored (Chapter 20).

Of course, the 30-day limit also applies on your end. You can cancel your contract if you have second thoughts about an assignment, make other plans, or just change your mind. However, if you cancel within 30 days, you risk financial penalties.

Where?

To increase the chances of finding a timely, desirable assignment, you need to acquire and maintain at least a few state licenses. A license in hand puts you at the top of your agent's list. As soon as the hospital completes privileging, you're ready to go.

At one point, I carried ten state licenses, all with different renewal days, DEA numbers, continuing medical education (CME) requirements (Chapter 17), as well as a multitude of CSPs, prescription monitoring programs, and various newsletters and password updates to keep up with. Hello Excel spreadsheets (Chapter 15)! It takes time to properly maintain numerous licenses, all of which is uncompensated.

The new Interstate Medical Licensing Compact eases the state license application process in some cases. However, most states don't participate, and it has severe limitations and additional cost (Chapter 9).

Clinical Work

Once the paperwork is complete (Chapter 9) and you finally arrive at work, you'll likely pose for a photo for your identification card, sign for a parking sticker, obtain multiple computer passwords, pee in a cup at employee health, and possibly submit to additional vaccinations, a tuberculosis test and an annual flu shot. You will work in a new environment where you probably don't know anyone, don't know where anything is, don't know whom to call for consults, and can't find the radiology department. Add an unfamiliar electronic medical record (EMR) (Chapter 16) and you've got the perfect first day!

One locum tenens physician advised, "If you are thrown in 'cold,' that is a distinct disadvantage and needs to be considered." Hear, hear!

A locum tenens physician must rapidly adapt to unfamiliar work environments that include new people, different protocols and equipment, and a variety of EMR systems. You must have a flexible and adaptable personality and be a quick study. If you don't possess these qualities, please stop reading and get back to your permanent job!

I worked at one county hospital where my predecessor stayed for only a few hours. She fled after orientation and never came back…

Compensation?

As a locum tenens employee, you will typically be paid by the hour (Chapter 13). The hourly rate is negotiable and depends upon many factors such as the urgency of the employer's need and availability of physicians to fill the role. It's old-fashioned supply and demand. If you are a top candidate, your experience and reputation may provide some leverage when negotiating the salary. Whether you accept the position is completely up to you.

You will not benefit from a CME allowance, company sponsored retirement plan, dental, disability, health, life, unemployment or worker's compensation insurance, sick days, vacation time, or other standard employee benefits. If you want any of these perks, you must pay for them yourself.

In addition, as an internal revenue service (IRS) 1099-MISC employee, you will not be subject to federal tax withholding, which means you must sequester part of your paycheck to pay IRS quarterly tax estimates. For example, let's say the staffing agency deposits the tidy sum of $15,000 into your bank account for two weeks work. That's pretty exciting until you realize that Uncle Sam already owns a sizeable portion of your paycheck. And don't forget that you'll have to pay health benefits and make a retirement contribution from what's left of the $15,000 after taxes. Consequently, you must exercise considerable discipline with your spending.

In some situations, you will be a salaried employee, but typically without benefits. In this case, you will be paid on a W-2 form and taxes may be withheld (Chapter 14). This situation is more likely on longer assignments.

You may need to pay state income tax to the state where you work. If your permanent residence is located in another state, you may have to pay tax there as well. To avoid double taxation, the tax paid in your work state will be credited to your home state.

For example, in 2017, I filed three state tax returns, Arizona and Minnesota, where I worked, and Rhode Island, where I was a resident. These were in addition to my federal tax return, 49 pages in all!

Are you looking for your CPA's phone number yet? You should be…

Appreciation?

You may certainly be appreciated by your coworkers if your presence improves their workload. You are also the "new guy/girl," and relatively vulnerable to pathological personalities. (Every hospital has them.) If you are unlucky enough that one of your patients experiences a "medical misadventure" on your watch, you may discover you are left high and dry with no supporters to testify to the high quality of your work. By definition, locum tenens physicians are "placeholders" without much of a support system. If things go bad, consider yourself a star in the movie, "The Expendables."

Local Politics?

Your fleeting presence allows you to ignore local politics without penalty. However, if you are one of those physicians who can't resist trying to improve ineffective policies and outmoded protocols, you may find yourself frustrated. As a transient physician, you will not be eligible for hospital committees and will have minimal if any ability to influence your work environment.

Limited Social Life

How much you socialize in the workplace depends a lot on your personality. However, because you are temporary, your colleagues may not take much personal interest in you. They know you will soon be gone. Not only won't you fit in to the social scene at work, but you've probably left your family and friends behind, so you're alone in a new city with only your stethoscope to keep you company. To combat loneliness on assignments, one locum tenens physician recommended bringing a pet (Chapter 20).

Chief Cook and Bottle Washer

Perhaps the foremost disadvantage of locum tenens is that you are a "one man (woman) band." You, and you alone, are responsible for finding an assignment, obtaining a state license and hospital privileges, traveling to the site, performing at the job, tracking expenses, paying federal and state quarterly tax estimates, arranging malpractice insurance, planning retirement, budgeting and tracking CME, scheduling future assignments, feeding the dog, and everything else.

You will likely need to hire an enthusiastic tax accountant to track your IRS 1099-MISCs from each assignment, file your state and federal tax returns, divine which of your many expenditures pass IRS muster as legitimate business expenses, and assist with financial planning. You also need a system to keep up with postal mail when you are away to insure that bills don't go unpaid, licenses don't lapse, and checks get cashed (Chapter 15).

More Trouble than it's Worth?

While the idea of taking months off from work every year may enthrall you, the administrative effort necessary to schedule, maintain state licenses, and run your own business will be considerable. You need to make an honest assessment of your organizational skills to determine whether you are up to the diverse demands of self-employment. Decide whether the freedom to work when and where you wish outweighs the ongoing hassle.

Many of us work not merely out of love for our profession, but because the bills keep coming. Although the hourly salary of locums may seem attractive, make sure you include uncompensated hours and lack of benefits in your income calculations. If your current lifestyle requires a full-time income, the predictable schedule of a conventional job may turn out to be more time and cost-efficient than back-to-back locum tenens assignments. Once you do the math, a full-time job with a steady paycheck, benefits, and no travel time may start to look pretty good!

Recommendations

Locum tenens work offers unmatched flexibility, but there's a high administrative price to pay. Take a cold, hard look at your balance sheet, innate flexibility and willingness to travel. That analysis will either steer you towards locum tenens or boost your enthusiasm for a traditional full-time job!

CHAPTER 5

Locums After Residency

The Decline of Private Practice

Physician-owned practices are in decline. Private practitioners face increasing government regulation, frequently changing reimbursement formulas, competition for insurance panels, dependence on cumbersome electronic medical record (EMR) systems, costly staffing and other overhead. These hurdles are particularly problematic for a solo practitioner. While not yet an endangered species, by 2014 the percentage of solo practitioners had dropped to 17% (Terry 2017).

An American Medical Association (AMA) survey revealed that in 2016, for the very first time, less than half of physicians owned equity in their practices (Terry 2017). The likelihood that newly trained physicians will hang up a shingle or partner in a group practice has been decreasing. In the AMA survey, physicians less than 40 years old are only half as likely to own their practices as physicians 55 years old or more.

The complex business challenges of private practice offer limited appeal to the majority of recent medical graduates, particularly when coupled with diminished income expectations. Millennial physicians joining the workforce are more likely than any previous generation to choose employed positions. These physicians may start out with locum tenens assignments in order to try different practice styles and geographic locations.

Job Interviews

I recently discussed the interview process for new residents with the Chair of Neurology at an academic medical center in New England. He reinforced my impression that millennial physicians are much more focused on work/life balance than prior generations. We reminisced that in our day, just the mention of work/life balance during an interview was the "kiss of death" regarding acceptance. Now it's expected.

These days, academic program directors must integrate a reasonable work/life balance into residency training programs if they hope to attract the best talent. Private practices and hospitals also recognize that employment packages must accommodate the large number of modern physicians who will not accept schedules that unduly sacrifice family time and personal endeavors.

Finding Your Way

After completion of a demanding residency and possibly a fellowship, young physicians may wish to work in one or more temporary positions before settling down. The demands of residency training leave little time to explore career options. Unless a resident has a clear vision of his or her immediate future, it may be difficult to choose amongst diverse job opportunities.

I have met several residents who held a firm grip on their future plans. For example, one married radiology resident planned to stay in town and join a local radiology group. He and his wife hoped to have children soon. As both of their families lived nearby, the in-laws could assist with child care. He had it all worked out.

Another resident planned to do a movement disorders fellowship at a tertiary hospital where he had previously worked as a research assistant. He liked the institution and loved the city. After the fellowship, he hoped to stay on as faculty.

Unlike these exceptional examples, the vast majority of trainees are still feeling their way. Choices after residency may be overwhelming: Research vs. clinical? Academic medicine vs. private practice? Group vs. solo? Government? Industry? East Coast vs. West? Big city vs. small?

Short locum tenens stints at different hospitals in a variety of locations will deliver a wealth of experiences, both good and bad, that provide a broad foundation for a young physician to arrive at a thoughtful permanent employment selection. Locum tenens experience may improve the success rate of a first job, which is only about 50%.

Changing Demographic

The number of female physicians is growing twice as fast as male physicians (Young et al. 2015). Dual physician couples are also more common. These demographic groups may find part-time employment with locum tenens an attractive option, at least until they clarify their professional goals in light of "real world" experience.

In the case of dual physician couples who do not complete their training at the same time, locums offers a convenient solution for one physician to start work while the other completes residency training or fellowship. This approach eliminates the need to relocate until both physicians are ready.

As a bonus, part-time physicians endorsed greater satisfaction, work control and less burnout than full-time physicians in one study (Mechaber et al. 2008). Newly trained physicians may also appreciate a break from intense years of medical school and postgraduate training they have just left behind.

Unlimited Parental Leave

In the U.S., the Family and Medical Leave Act of 1993 guarantees eligible employees 12 unpaid weeks. Three states, California, New Jersey, and Rhode Island, insist on paid family leave. Despite these regulations, many physicians don't even take their own recommendations for parental leave (Lenhart 2018). They fear abandoning their patients, burdening colleagues with additional patient care or administrative duties, losing their position or other consequences (Welch 2017).

A mother's early return to work after childbirth can disrupt attempts at breast feeding as well as bonding with a cherished new family member. New moms may be torn between conflicting desires to stay home with their child or return to work. "Mommy guilt" can affect fathers as well (Pinola 2015).

Locum tenens offers a solution. Physicians can take as much time off as they wish between assignments. There is no commitment (or expectation) to resume work. When it comes time to return to patient care, new parents can choose short assignments in order to continue to balance family responsibilities.

Further, it's really true that "when you're off, you're off." Between assignments, there are no work obligations. No emails, phone calls, patient communications or administrative responsibilities. None.

For young physicians starting a family, locums offers the opportunity of improved work/life balance. As a locum tenens physician, you may provide coverage for other doctors taking maternity or paternity leave. When it's your turn to start a family, employment as a locum tenens physician allows you to schedule as much leave as you need, albeit unpaid.

As a caution, more than two years outside clinical practice can cause insurmountable hurdles when attempting to return (Chapter 9). The solution? If a parent chooses to spend several years as a full-time Mom or Dad, brief locum tenens assignments can keep skills and references fresh.

Gaining Experience and Financial Footing

Locum tenens is a practical tool to prepare for private practice. One young female physician wrote, "I am family medicine trained and have been working locums for two years to pay off loans and get enough money to start my own family medicine practice. I will likely keep doing locums after I open my practice to keep up my emergency skills and bring in the extra money."

Compensation

Recently graduated residents accustomed to a trainee's salary should find a locum's salary more than adequate. As compensation is directly proportional to hours worked, newly hatched physicians can decide how much they want to earn and schedule assignments accordingly.

Shift Work's OK

Senior physicians often denigrate "shift work" as unworthy of the high calling of the practice of medicine. Many of these doctors provided a lifetime of 24/7 access to their patients. They took vacations reluctantly, if at all.

Times have changed. Few contemporary physicians accept 24/7 on-call responsibilities day in and day out. (I'm happy to do it every other week...)

Since 1989, when New York State imposed work hour restrictions for hospital residents, physicians-in-training have become accustomed to mandatory shift work rules. In 2011, the Accreditation Council for Graduate Medical Education (ACGME) imposed an 80 hour/week work limit and a day off every week (Temple 2014). Shift work inherent in locum tenens isn't likely to present a philosophical barrier for younger physicians.

Ask Questions

With any new job, it's essential to know everything that's expected of you. As a locum tenens physician, you may find yourself in a new facility with little supervision and minimal guidance regarding daily duties. Your employer expects that you'll figure it out and hit the ground running.

An orthopedic surgeon who works locum tenens advised, "Ask a lot of questions at the beginning and get very clear answers regarding your responsibilities. Be clear on whom to contact with any questions or concerns in the various facts of the job: medical documentation, EMR, ER responsibility, consults, transfer of patients, liability, etc."

One Caution

One locum tenens anesthesiologist cautioned residents from considering locums as their first clinical job:

> You have to be comfortable in your skills. Most residents nowadays are undertrained, and if you are not comfortable taking care of sick people, I don't think locums is for you. Locums is kind of a clinical Siberia, everyone expects you to be able to do your job and be proficient. There are a few rare residents that can just jump in and be comfortable in any clinical situation, but I wasn't like that.

It's true that most locum tenens physicians are more experienced than newly graduated residents (Chapter 6). Before you sign on to a locums job, take a breath, sit down, and assess your skill level and adaptability. As much as your locums agent may encourage you to take the job, it's better not to sign on if it's going to stress you beyond your comfort zone. Only you can make that decision. Sage advice from the Temple of Apollo still holds, "Know Thyself."

Recommendations

Newly trained physicians confident in their skills who need to pay off student loans, save money to start a private practice, or have not yet decided on their practice style and location, should consider locum tenens.

CHAPTER 6

Locums Mid-Career

Six Big Reasons for Locum Tenens

According to a recent survey, first-time locum tenens physicians are most likely to be mid-career (49.2%), followed by retired physicians (35.9%) and those right after residency (14.9%) (Staffcare 2017).

Mid-career physicians have been in practice approximately ten years or more. They may be in group practice, hospital, solo or other permanent employment. They've got a predictable routine and work hard. They've acquired solid clinical skills, as well as an assortment of assets and responsibilities. They earn a respectable income. For most, life is good.

However, there are many reasons why a mid-career physician may consider locum tenens. Here are six important ones:

- Extra income
- Travel
- Wait out a non-compete clause (restrictive covenant)
- Burnout
- Bridge to another career
- Family time

Extra Income

Physicians are generally considered "highly compensated" individuals, typically earning six-figure salaries. According to the U.S. Census Bureau, a salary of $214,462 or more lands one in the coveted category of top 5%

income earners in the U.S. (Lake 2016). However, even such a substantial salary may come up short when faced with the pricey demands of modern life. In a high-rent district like Silicon Valley, $250,000 a year may lead to one "barely scraping by" (Roitman 2017).

Many mid-career physicians add locum tenens shifts to their permanent jobs in order to supplement income. Extra money may be needed to invest in personal or commercial real estate, pay off a mortgage, add to a retirement account, pay college or private school tuition, recover financially after an illness, repair finances after a divorce, front a loan to a needy family member, take an expensive vacation, fund an extravagant purchase such as a luxury vehicle, aircraft or yacht, or other purpose. Physicians are also highly susceptible to "lifestyle creep," and may require additional income simply to balance the family budget (White Coat Investor 2016).

Employed physicians may participate in locums as long as the work occurs during off-hours or vacation and does not violate non-compete or other contractual constraints. The temporary nature of locum tenens makes it perfect for "moonlighting" opportunities.

Travel

Traditional medical practice can prove an insurmountable obstacle to exotic travel adventures. During my eight years working in a busy private practice neurology group, a "long vacation" consisted of two weeks that had to be scheduled six months in advance.

Like me, many mid-career physicians put their dreams of travel on hold as they soldiered through medical school, residency, fellowship, and struggled to stabilize a fledgling medical practice. A locums career can free up weeks or months for extended trips, unimaginable for physicians in conventional practice. In addition, income from locum tenens assignments can underwrite travel expenses that might otherwise be out of budget.

After my first trip to the island of Borneo in Southeast Asia, I discovered, much to my surprise, that my extravagantly long two-week vacation resulted in minimal disruption to my patients' lives. My partners competently handled all the emergencies and urgent follow-ups. When I returned to the office, it was as if I had never left, except for the massive pile of charts on my desk awaiting signatures and review.

In those paper chart days, patients expected their physicians to be *always* available. But somehow, with the kind assistance of my partners, all my patients survived without me! This was a lesson well-learned.

A subsequent career of full-time locums has allowed me to periodically put the rat race on hold and explore Southeast Asia for months at a time. (I'm writing this chapter on my wife's iPad in our little house in the Philippines...) As a locum tenens physician, I don't get paid when I'm not working, but when the palm trees are swaying, and the sand feels warm under my feet, it doesn't seem to matter much.

Non-compete Clause

Also referred to as a restrictive covenant, a non-compete clause is a contractual obligation that prohibits a physician from leaving an employer and practicing in the same community for a stipulated period of time. The restricted geographic area and duration vary according to the contract.

One 62-year-old physician found himself mired in his medical group with an overwhelming workload and underwhelming salary (Chapter 20). His financial future appeared bleak as the group careened towards financial ruin. He loved living in the Pacific Northwest and didn't want to move, but his contract's non-compete clause prevented him from working with any local hospitals or medical practices. He was stuck.

For months he suffered long hours in the clinic, read hundreds of electroencephalograms, and watched helplessly as his paycheck shrank. Finally, he reluctantly returned an unsolicited phone call from a locum tenens recruiter. After serious due diligence, he signed on for his first ever locum tenens job. Over the next two years, he completed several out-of-town assignments.

As a locum tenens physician, he alleviated his misery, earned a respectable income, and had time to interview for several desirable permanent positions. He enjoyed the novelty of locums assignments, which included both teaching and private practice positions. These diverse experiences lent insights into the pros and cons of various work environments, improving his ability to assess future employment options.

After two refreshing years of locums, the restrictive covenant expired. He was free to resume work in his community. In the meantime, he had received a job offer from across the country that was so attractive he considered moving. Without locum experiences, he never would have learned about the new job or had the courage to pursue it. The last time we spoke, his biggest stressor was whether he should stay put or relocate to the East Coast!

Family

Most physicians are dedicated to their patients and work more than eight hours a day. It's no surprise that mid-career physicians often discover that work dominates their existence with little time left for family and friends.

Locums offers a solution, although it may seem paradoxical. A week or two away from home can be followed by a similar period at home without work obligations. Some locums physicians find this strategy delivers more "quality time."

A number of my assignments have been 7 days on/7 days off, allowing alternating immersion in clinical work and personal pursuits. On the job, I'm 100% focused on patient care. When I'm off, I'm off. It works for me.

For example, for five months last year, I commuted weekly from Phoenix, AZ, to Minneapolis, MN. It dawned on me that I was a frequent traveler when I recognized the flight attendants!

A pulmonary critical care physician recently opting for *The Locum Life* told me, "I have a three year old. When I am working in the intensive care unit, I am basically not home during her waking hours. I am starting locums to spend more time with my family. At least that is the theory. I will be able to update you in about a year."

Some locum tenens physicians take their families with them on extended assignments. A reasonable schedule and lack of administrative responsibilities permits a "working vacation." Such a strategy may help prevent burnout (see below).

Burnout

Burnout is characterized by a loss of energy, enthusiasm, and efficacy at work (Valcour 2018). Much has been written about "physician burnout," and none of it is encouraging (Bernat 2017). Administrative hassles, unrealistic high workloads, inadequate support, and frustrating electronic medical record systems have become the norm for many physicians.

While individuals bear some responsibility for combating burnout, they may be overwhelmed by factors beyond their control. Locum tenens offers opportunities to break with an established routine and oppressive "toxic" environments. A change of venue may provide a welcome break from the daily grind. It may also offer insights into better alternatives for the future.

One global locum tenens agency proposed the following:

> *Instead of seeing job burnout solely as a problem, see it as an opportunity to change directions. Tired of hearing about physician burnout? Here's one possible solution: revitalize your medical career with an international locum tenens opportunity.*

If the hectic pace of hospital medicine has you exhausted, you may want to try a clinic job. Conversely, if your clinic patients' problems leave you bleary-eyed by the end of the day, perhaps you'd like to revisit the urgency of hospital consults. As long as you have the proper skills, you should be able to find a locum tenens position that adds another dimension to your routine.

A new position at another hospital will also expose you to a whole new set of practitioners. Each hospital has its own culture. Once you see how the other half lives, you may discover strategies to improve your own environment. Alternatively, a brief stint at another institution may reveal just how lucky you are. Either way, it's rare to come away from an assignment without having gleaned at least a few clinical pearls and administrative lessons that can be applied to your permanent job.

Women are particularly susceptible to burnout (Peckham 2018). A mid-career physician recently telephoned me to discuss the possibility of practicing locum tenens. It was easy to see that she suffered from burnout. In her own words:

> *I've been at this for 12 years, and the way I think things should be run has nothing to do with the way the big multispecialty group runs. Right now I am paid very well, but I don't really need that. I am less worried about the money than I am about my sanity and my quality of life. When I get home, I'm so exhausted and so burned out. I don't even have the energy to tuck my children into bed. What I want is more flexibility and more freedom.*

You can imagine what I told her. She needed to take a break and think through her long-term options. A short locum tenens assignment was just what the doctor ordered.

This unhappy physician displayed several qualities necessary for success with locums. She was highly motivated, confident in her clinical skills, and ready for a change. From a practical point of view, she had the financial security to explore new options and her physician-husband's employer could cover her health insurance. She was well positioned to try something new.

What remains to be seen is whether she has the adaptive social skills to jump into a new environment and thrive. I suggested she find a good staffing agent, try an assignment for a week or two, and see what happens.

Stigma

This frustrated physician also told me she had been thinking about locums, but had reservations regarding how it might affect her long-term career. "I trained at an ivory tower, Yale," she explained, "and I thought that locum tenens was for someone who couldn't hold down a job."

I've asked a number of physicians whether they experienced stigma associated with locum tenens. It's fair to say that if your career goal is chief of neurology at the Mayo Clinic or Massachusetts General Hospital, bouncing around from various locum gigs is not likely to get you there. Outside of academics, however, the proof of the pudding is in the eating, which in clinical medicine translates to high quality care and getting along with peers and staff.

According to an article in the *Nevada Daily Mail*, "At one time, locum tenens physicians were viewed as second-rate doctors who could not hold a permanent position. Today, such positions are becoming coveted, attracting high quality physicians" (Brann 2017).

Upon arrival at a new assignment, I have sometimes encountered a raised eyebrow when I introduced myself as a locum tenens physician. However, my competence has not been questioned. Treatment as a second class citizen never occurred.

A recent Staffcare survey revealed that 94% of health care facility managers used locum tenens in the last year, the highest percentage ever recorded. If a stigma exists, it's not affecting the job market (Staffcare 2017).

A follow-up question my burned out physician colleague asked is whether time in locums could thwart one's chances for a full-time position. It may. I know of two physicians who practiced locums for more than a decade and experienced pushback when they finally applied for permanent jobs. The employers worried that the peripatetic physicians wouldn't stay put, probably a reasonable caution on their part.

Physician recruitment is time consuming and expensive for the hiring group or hospital. The last thing employers want is to on-board a new employee, with all the administrative work and expense that entails, only to have the physician leave a few months later.

When I recently applied for a full-time faculty position, I had to address similar concerns. I reassured the employer that my varied locum experiences had not only honed my clinical skills and improved my abilities to work with a variety of colleagues, but also showed me the type of professional environment that would facilitate my development into the "best physician I can be." In addition, my life circumstances had changed, and I wanted to settle down. Who could resist such a candid and truthful explanation?

Bridge to a New Career

For a mid-career physician, locum tenens can serve as a bridge to a new career. After ten, twenty or more years in clinical practice, some physicians wish to further their education with an MBA, move to hospital administration, the pharmaceutical industry, or take on nonmedical roles, such as real estate magnate, restauranteur, or full-time blogger. In preparation for a new job that may result in an initial salary dip, locums assignments can bolster savings.

The practice of locum tenens also allows physicians to maintain their clinical identity, skills and qualifications, providing a nice fall-back position should the new vocation not work out. With locums, you don't have to burn any bridges while you test the waters of a new career.

Remember, it may be difficult if not impossible to find employment as a clinician if you have been away from patient care for more than two years. Even if your new nonclinical job is going great, occasional locum tenens assignments reset the clock and preserve your option to return to clinical medicine.

Rewards of Locum Tenens

For mid-career physicians, occasional locum tenens assignments supplement income without surrendering the benefits of a permanent position. Locums offers an alternative to conventional practice and opens doors to new and varied clinical and social experiences. Locum tenens may also smooth the transition to a second career or even become a new lifestyle. Physicians motivated to try locum tenens will likely find their efforts amply rewarded, both emotionally and financially.

Recommendations

Mid-career physicians should not hesitate to pick up the phone and discuss their thoughts with a staffing agent at a National Association of Locum Tenens Organizations (NALTO) registered locum tenens company (Chapter 8). Sometimes, just talking with an agent will clarify whether you are a good candidate for *The Locum Life*.

CHAPTER 7

Locums Late-Career

Who is "Late-Career"?

"Late-career" physicians are in their last third of professional practice. Since physician practices vary in length, late-career cannot be defined by years in practice. What is late-career for one physician might be mid-career for another.

Similarly, no specific chronological age defines a late-career physician. While commercial airline pilots must retire by age 65, physicians do not face a mandatory retirement age. In 2016, nearly 100,000 actively licensed physicians aged 70 or older practiced in the US (Burling 2017).

Unlike professional athletes, physicians blessed with good health can practice commendably and continue to enhance their skills well into advanced age. While physical abilities may diminish with age, compassion, resilience, and wisdom may increase.

Why Locum Tenens?

Late-career physicians may choose locum tenens for any of the same reasons as mid-career physicians (Chapter 6). Extra income, travel, waiting out a restrictive covenant, burnout, bridge to another career and more family time may all prod late-career physicians to try locum tenens.

Locum tenens is particularly popular with mature physicians. Three quarters of locum tenens physicians are older than 50. In fact, a third of locum tenens physicians waited until after retirement before trying locums (Staffcare 2017).

Challenges of Retirement

As late-career physicians observe their colleagues getting younger and younger, they begin to consider the prospect of retirement. However, a "disproportionate number" of physicians defer retirement until after the traditional age of 65 (Silver et al. 2016). Many are understandably reluctant to abandon a complex skill set cultivated over many years and wish to continue working as long as they are able. Many experienced physicians have become invaluable medical resources. Who hasn't sought the sagacious advice of a slightly hunched, silver haired colleague draped in a long white coat?

While retirement may seem to some like the glorious culmination of a successful career, it presents challenges of its own. Maintaining financial independence and a sense of purpose are two common and critical considerations for retired physicians.

Income

Like other professionals, some physicians retire as soon as they achieve financial independence. On the other hand, many physicians who no longer require a yearly salary continue to work.

For some physicians otherwise ready to retire, financial independence may prove elusive. Chronically unemployed progeny, expensive school tuition, ill-considered or unlucky investments, medical illness, one or more divorces or any number of other financial calamities may undercut retirement savings. Social security disbursements are not likely to satisfy most physicians as their sole source of retirement revenue. Locum tenens assignments may help close the gap.

Some retired physicians discover that they need to resume medical practice in order to supplement their savings. For example, during the recession of 2008, many physicians saw retirement savings drop by 30 percent or more. In a single day, comfortable nest eggs turned into omelettes.

Physicians may wish to return to work for reasons other than income (see below). The existence of a variety of physician re-entry programs testifies to the phenomenon of physicians leaving retirement to rejoin the workforce (physicianreentry.org).

A Sense of Purpose

Despite having achieved financial security, many late-career physicians are reluctant to retire. While contemporaries in other careers developed hobbies, these physicians eschewed extracurricular activities in order to devote their full energies to a demanding profession.

For these dedicated individuals, abrupt transition into a retirement with unlimited free time may result in bottomless boredom rather than boundless bliss. Long days spent on manicured golf courses, lingering evenings at swanky restaurants, and protracted vacations cavorting on cruise ships may fail to engage physicians accustomed to a daily barrage of life and death decisions. The pleasure of leisure activities cannot match the value they attach to their professional contributions. Loss of identity as healers may leave them feeling adrift without purpose.

According to a recent CompHealth survey, enjoyment of the practice of medicine, social aspects of working, and desire to maintain an existing lifestyle were the three most common reasons physicians planned to work past age 65 (Saley 2017). While physicians may not be as satisfied with their careers as they used to be, 82% stated they were either satisfied or completely satisfied. Surgeons expressed the highest job satisfaction of any medical specialty and were the least likely to look forward to retirement.

One retired pediatric surgeon, a former chief of surgery with five decades of experience, insisted that he never intended to practice medicine after retirement (ReachMD.com). However, less than a year after stripping off his last pair of sterile gloves, he realized he was "wasting myself" when he could be "really making a difference in the life of children." Still physically and mentally fit, his productive days did not have to end.

This doctor discovered that nothing in retirement could match the satisfaction of helping sick children. However, he did not want to resume the demanding responsibilities of a full-time surgeon.

Locum tenens assignments empowered him to practice surgery on his terms. He was able to continue to provide specialized medical care for children, which had always been his passion. As a bonus, he enjoyed the role of mentor to young surgeons at different facilities. He enhanced his legacy by teaching skills and sharing enthusiasm with a new generation of pediatric surgeons. Retirement couldn't compete with that!

Age Discrimination

Retired physicians may find it difficult to find a new job for a variety of reasons. Their skill set may be rusty, requiring retraining. They may be reluctant to relocate. Computerized medical record systems may intimidate. The prospect of assuming a "junior" status in an organization may seem demeaning.

Age discrimination may be another factor that thwarts job searches and prevents mature physicians from securing permanent employment. Although there are legal protections against age discrimination in hiring, such bias likely still exists. When employers assess applicants for full-time, permanent positions, they consider growth, leadership potential and long-term viability as well as suitability for current job requirements. Older physicians coming out of retirement may not appear a wise investment.

Locum tenens employers, on the other hand, face an immediate need for clinical expertise. Older physicians represent a wealth of knowledge, experience and an invaluable resource. Long-term employment considerations don't come into play. As long as retired physicians have kept up their continued medical education, licenses, and skills, they should have no problem finding locum tenens assignments and succeeding on the job. The lack of age discrimination in the locum tenens industry likely contributes to the popularity of locum tenens amongst mature physicians.

Graceful Exit Strategy

Senior physicians may wish to continue to work but may no longer possess sufficient stamina or enthusiasm for full-time practice. Excessive workload is a major reason that forces physicians into retirement (Silver et al. 2016).

Many physicians wish to keep working part-time even after "retirement" (Saley 2017). However, medical groups may not offer a truncated schedule. Younger members may resent older physicians taking days off, less call, or leaving clinic early. They may view semi-retired physicians as less productive, consuming clinic resources but adding little to the bottom line. Because of these perceptions, half-time work often translates into less than half-time compensation. Older physicians may see no option but to continue full-time or retire completely.

Locum tenens offers an attractive alternative to the binary choice of full-time work vs. total retirement. A lifetime of experience and polished skills constitute an excellent combination for gratifying locum tenens experiences and a graceful transition from full-time practice to retirement.

Recommendations

Late-career physicians have become experts at assessing clinical scenarios, but may feel locked into demanding full-time practices that no longer match their priorities. Discuss desires, dreams and needs with a financial advisor as well as those close to you. A follow up chat with a staffing agent ought to reveal whether locums is an appropriate use of your time.

CHAPTER 8

Choosing an Agency

Staffing Agencies

More than a 100 staffing agencies link physicians with locum tenens employment opportunities (Appendix 2). I've used two of the largest, CompHealth and Staffcare. CompHealth, established in 1979, is the oldest locum tenens company (Chapter 2). Staffcare, a subsidiary of AMN Healthcare, is another venerable outfit. About 80% of locum tenens physicians depend upon staffing agencies to facilitate job searches and administer contracts, transportation, housing and paychecks.

Both CompHealth and Staffcare belong to the National Association of Locum Tenens Organizations (NALTO). Established in 2001, NALTO holds locum tenens agencies to practice and procedure standards as well as an ethics code. These principles promote appropriate business practices. NALTO's board of directors consists of locum tenens industry professionals. At last count, more than 70 locum tenens agencies belonged to NALTO. A member list appears on the NALTO webpage (www.nalto.org).

Like Buying a House

A locum tenens agent works in much the same way as a real estate agent. Real estate agents pair home buyers to home sellers; locum tenens agents match physicians with temporary job opportunities (Chapter 1). Both types of agents earn commissions, which incentivize their efforts. Theoretically, the cost of commissions could raise house prices and lower physician compensation, respectively. In practice, it may not work out that way.

Advantages of Working with an Agent

Agents are always looking for new physicians to represent and are eager to talk to you. In an effort to find more physician clients, locum tenens companies sometimes sponsor booths at national medical meetings. This is a great opportunity to talk to agents in a low-pressure environment and ask questions.

In lieu of a face-to-face meeting, you can browse agency websites. Appendix 2 offers a long list of active agencies. Pick a few and check out their assignments. If you see one you like, call and discuss the opportunity with the staffing agent. Alternatively, if there's an online form, complete it and an agent will quickly contact you.

When you speak with agents, they will want to learn your medical qualifications, work history, goals and availability. Here is a list of questions my CompHealth agent sent me.

- Are you board certified? Is so, when do your boards expire? If not board certified, are you scheduled to take your boards? If so, when?
- Are you fellowship trained? If so, what is your subspecialty?
- Please verify the best phone # and whether or not this is a cell/office #, email address and your preferred mailing address to make sure what we have on file is accurate.
- Where are you actively licensed?
- Do you have any expired licenses?
- Do you have any licenses in process?
- Are there preferred states you would like to work in if we can get you licensed?
- Have you worked in an outpatient setting in the last 24 months?
- Have you worked in an inpatient setting in the last 24 months?
- Do you prefer inpatient or outpatient work?
- What patient population do you see? Adults, pediatrics or both?
- Do you have an active unrestricted DEA?
- Has a malpractice claim ever been filed against you? If so, how many? What were the payouts?
- Any issues on the National Practitioners Data Bank or the Federal State Medical Board?

- What is your availability? Examples: full-time, 1-2 week stints at a time, a few weeks per year or weekends?
- When was your last patient contact?

As you can see, the questions are comprehensive. If your answers are satisfactory, the agent will send you a formal application.

I reviewed my files and found the initial paperwork I submitted to Weatherby Healthcare a few years ago. The 12 page application requested biographical information regarding education and residency training, work history, hospital affiliations, licenses, DEA numbers, and four references. There was a page and a half of obligatory "yes/no" questions regarding any "actions, limits and sanctions" and "disciplinary action information." The application also requested authorization for release of information.

Digging up all this information may take several hours. Scan and save documents (Chapter 15), as you will need them for future hospital privilege and license applications (Chapter 9).

Once you have registered, the agent can get to work to find you a position. A good agent can assess whether your candidacy is appropriate for a particular employer. Many assignments are recurrent, and the agent may have worked with that employer before and knows what to expect. Your agent wants you to have a good experience so you'll return for future assignments. Of course, the agent also needs to please the client who's paying the bill.

When an agent calls with an opportunity, respond immediately. Agents work with multiple qualified candidates. The first one to say "yes" gets the job.

If the agent thinks you are a good match, and the timeframe is reasonable (see below), he or she will do everything possible to place you. The agent's responsibilities include contracting, assisting with hospital credentialing, state licensing if needed, travel and housing arrangements, reimbursement for work-related expenses, and acting as a go-between should problems arise during negotiations or on the job.

Staffing agents specialize in a few medical specialities and learn the marketplace. Just as a real estate agent can accurately price a house in a specific neighborhood, staffing agents know the going rate for the medical specialties they handle and stay on the lookout for upcoming opportunities.

If you ask, an agent may connect you with a physician who has previously worked at the same facility. There's nothing more valuable than speaking peer-to-peer with someone who's been there before.

I recently had the opportunity to repeat a neurohospitalist assignment. On the first go-round, I covered the stroke service in a teaching hospital while the administration recruited a permanent faculty member. My days were spent rounding with residents, hurrying to the ER for stroke codes, and treating the occasional homeless patient who showed up in clinic. The job was so much fun I told the chief of neurology that I would like to stay on permanently.* He agreed, since I had already proven myself. Unfortunately, the powers that be at the medical school had already chosen their preferred candidate. I never got the position.

Two years later, the same hospital asked me to return. Why? Seems like the doctor they hired (instead of me) had come and gone.

Why were they contacting me? The hospital had already hired a new replacement (instead of me), but the physician was still in fellowship and couldn't start for another six months.

Admittedly, I was miffed they had twice exercised such poor judgement regarding my application for the permanent post. But it was such a great gig I swallowed my pride and agreed to work until the new faculty arrived. But I did suggest a raise (see below)!

Although the hospital contacted me directly for a second tour, it was bound to work with the same staffing agency because they had signed a two year contract. I dutifully communicated the job offer to my agent, and she picked up negotiations from there.

One or two year contracts between facilities and agencies are standard. These agreements discourage employers from "poaching" doctors from agencies. Without such legal safeguards, once a candidate arrived on site, the employer could terminate its agency contract, rehire the candidate at the same rate (or less), and pocket the agency's commission. Similarly, if the facility offered me a permanent contract during the two year time frame, the agency would be entitled to a recruitment commission.

But here's where my agent really demonstrated her value. Not only did she assist with credentialing paperwork, which unfortunately had expired from the first assignment, but she succeeded in negotiating higher compensation. The hospital had offered the same hourly rate that I received during my first assignment. Now that I was a more worldly locums doc, I realized that the pay had been on the low side. Plus, there had been two years of inflation! When I discussed this with my agent, she agreed that a boost in my hourly rate seemed reasonable. She took care of it, and spared me the awkwardness of asking for money.

I didn't have to say a word to anyone at the hospital, and a pleasantly revised contract appeared in my email. Thanks to the agent, everyone was happy; the hospital got the doctor they wanted, the doctor got the compensation he wanted, and the agent got a higher commission!

When beginning a search for a locum tenens assignment, I browse email job postings from several locum tenens companies about 4-6 months before my planned start date. Other than word of mouth, it would be difficult to obtain this information. No centralized database exists for locum tenens opportunities. (For that matter, the same holds true for permanent physician jobs.) Here's an actual locum tenens advertisement from Pacific Companies for a position in Napa Valley, California:

> *Outpatient position (clinic) but will need inpatient privileges as there will be some consults and possible rounding. Must have inpatient experience in last 24 months.*

> - *Could use full time locum but would like at least 2 weeks a month. Will need until we hire.*
> - *M-F 8:30-5 (40 hours a week)*
> - *No nights, no Call*
> - *Only Neuro in hospital as we are recruiting for a perm Neuro*
> - *Cerner EMR*
> - *Volume Expectations 14-18 a day*
> - *Things you will see-Epilepsy, headache/migraines (botox skills nice to have), Parkinson's disease, dementia/Alzheimer's, MS (must be familiar with MS meds), tremors, neuropathy, trigeminal neuralgia, Huntington's disease. No sleep disorders, no neuro-psych, no neuro-ophthalmology, no patients under 18. EEGs mandatory EMG and NCS not as much but nice.*

- *Required: Board Certified (ABMS approved) or Board Eligible (7 years from training).*
- *Required: Active California Medical license*

Agencies list opportunities by state. Breakdown by state is essential, not only to appeal to your geographic preferences, but because you must have the respective state license in order to work.

Obtaining a new state medical license entails a pile of paperwork and endless delays (Chapter 9). I know, because I've done it 13 times. Consequently, a license in hand puts you in a very favorable position from both the agent's and employer's points of view.

I search opportunities in states where I already have a license. Given the amount of effort necessary to obtain a new state medical license, I would rather skip that step. I would only go down that road again if it was an incredibly attractive position or my savings ran out.

I did have such an appealing opportunity a few years ago that required three new state licenses. It was a telemedicine job with a major academic institution. Despite the onerous state licensing procedure, there's no question I came out ahead. The job allowed me to develop valuable telemedicine skills, bestowed me with an academic appointment, and offered extraordinary compensation. After a year and a half, I handed over my headset to a new full-time employee.

What about the three licenses? I've already let one of them lapse, but I'm hanging on to the other two, Arizona and Nevada. These states often look for locum tenens physicians, and I like to keep my options open.

If you don't have the appropriate state license for an upcoming job, your agent can advise whether it makes sense to get one. An experienced staffing agent knows how long licensing takes in a particular state.

Your agent should discourage you from applying for an immediate need in Texas, for example, because this state is notoriously slow to grant licenses. Texas requires an online application, fingerprints for a criminal background check, and a medical jurisprudence exam. Amassing all the required materials to complete your application will consume weeks to months, and the board may take nearly two months before it processes your application. Licenses are only issued at two-week intervals. If the board detects anything irregular in your application, such as problems with "professionalism," your application heads to its own private purgatory from which it may never emerge.

Brief assignments of a few days or weeks that require a new state license are not worth the effort unless you plan to work in that state again. It's too much trouble to apply for a state license and maintain it for a single, short assignment. In addition, future applications for other state licenses will require records of all past licenses, adding to your paperwork burden.

Note that state medical boards bear no legal requirement to grant you a license. Legally, they are simply obligated to process your application. The upshot is that you, the applicant, are completely at their mercy.

In my imagination, I often hear the state employee at the other end of the phone, "Sir, if you are not nice and patient while I hold your future in my hands, your application might get misfiled."

Waiting for licenses has cost me tens of thousands of dollars in lost income due to bureaucratic delays. A single federal medical license anyone (Chapter 9)?

Agents are paid primarily on commission. The sooner you get to work, the longer you work, and the more you get paid, the better for them. Agents also thrive on repeat business. If you have a good experience, you are likely to want another assignment. You are under no obligation to use the same agent or agency, but it's the path of least resistance, so why not? Agents will endeavor to befriend you, become your confidant, and do whatever it takes to make you a satisfied and repeat customer.

Another advantage of working with an agency is a 24-hour travel concierge service. On one miserable trip, I had to call after midnight when my flight to Sioux Falls got cancelled. The travel agent rebooked the flight and notified the hospital that I was on my way. Even though I was unhappy at arriving late on my first day, at least the administration knew it wasn't my fault.

Finally, an important advantage of working with an agency is prompt payment. You provide the service, document your hours, and a direct deposit shows up in your bank account.

Large agencies like CompHealth and Staffcare use online software to track hours and generate your paycheck. To facilitate accurate and timely payments, log your hours daily. If you wait until the end of the week or month, you may forget overtime hours and miss out on hard-earned dollars.

If you don't work with an agency, you will get paid directly from the hospital or clinic. If there's a problem, it may be difficult to resolve on your own.

Disadvantages of Working with an Agent

An agent is a third party, and like any third party, it distances you from the client and adds a layer of complication. Every agency also takes a percentage of your salary in order to pay the agent's commission, cover its other expenses, and make a profit. Just like real estate agents, locum tenens agents prioritize closing the deal. An unscrupulous agent may put that goal ahead of your best interests.

For example, one agent strongly encouraged me to get a Mississippi license for an enticing job opportunity. The Mississippi medical board requires not only the usual mind-numbing application, but a mandatory in-person interview. My reluctant visit to Jackson, Mississippi, included an excruciating seminar detailing the various and sundry ways a physician could fall under investigation by the state medical board for professional misconduct. There was also a memorable visit to a nearby restaurant whose fried food had been meticulously optimized to encourage atherosclerosis.

To their credit, the agency paid the Mississippi application and travel expenses. However, by the time the painful licensing process was completed, the client had filled the position! An experienced agent would have known how long Mississippi licensing would take. This fool's errand consumed three full days, none of which was compensated.

Now I have a beautiful Mississippi license, suitable for framing, which I have never used. It was hard earned, so I regularly renew it in case I need it someday. As luck would have it, I've recently moved to Memphis, Tennessee, with the Mississippi border just a few miles away. Maybe that license will come in handy after all…

It's important to remember that staffing agents are not physicians. They may have little idea of what you actually do at your assignment. I've had agents tell me, "You're expected to see 15 patients a day," but they don't know whether that number refers to new patients, follow ups, or both. The answer makes a big difference. In neurology, new patient evaluations take much longer than follow-ups. Fifteen new patients would make for a miserable, demanding day. A mix of 15 new and follow ups, however, would be fine.

I have yet to speak with any agent who could provide this level of detail on the first phone call. But most agents diligently respond to questions. After all, their job is to find you a suitable assignment. If you don't get the answers you need, you'll switch to another agency. It's bad business for an agent to lose you as a client before you even start.

Last Minute Lifestyle

Locum tenens positions appear and disappear without warning. One of the reasons to register with more than one agency is to insure that when you are available to work, you will locate a suitable assignment. When an opportunity appears, you may have to drop what you are doing in order not to miss out.

In addition, agency contracts typically allow for cancellation by either party up to 30 days prior to the start of an assignment. This provides great flexibility for both you and the client clinic or hospital.

Flexible scheduling is one of my priorities, but these rules wreak havoc on long-term planning. Your potential employer is recruiting for a permanent replacement and regards you as a "placeholder" (Chapter 1). If recruitment succeeds ahead of schedule, you will be out of a job.

One hospital I was scheduled to work at cancelled at the last minute, right on the 30-day mark. I was left with an excess of free time and a gaping hole in my budget. Long hours at my desk filling out forms for credentialing, trip planning, and juggling my schedule had been wasted. Ideally, my agent should have given me a heads up that a cancellation was in the works, but maybe she didn't know. I never learned why the hospital cancelled the assignment. That's par for the course.

Free Agent

Even if you contract with a locum tenens agency, you may work simultaneously with other agencies. While it may seem disloyal to work with a competing agency, the lack of a centralized locum tenens job board demands it. Some job posts may be duplicated by rival agencies, but others may be unique to a particular staffing agency.

Consequently, you increase your chances of finding attractive and timely assignments if you play the field. While staffing agents relish fidelity, no moral or legal expectation of exclusivity exists between physicians and locum tenens agencies.

Automated Agencies

Several new locum tenens agencies, Lucidity, MyCoverDoctor, and Nomad, promise less hassle and higher physician payments by operating without agents. These automated agencies rely on computer technology rather than human beings to match candidates with employers. Physicians complete a website profile, then search posted positions. Salaries may be negotiated online.

Many physicians will recognize this approach from medical school, when they experienced something similar when preparing for "Match Day." While the absence of an agent may streamline the process and eliminate costly commissions, it also removes the opportunity for an experienced and sympathetic agent to listen and advise.

Automated locum tenens agencies are an interesting experiment. They follow in the mold of Airbnb, Lyft and Uber, which have eliminated the middleman while pairing individuals with needed services. I haven't had the opportunity to try this new approach, and it's too new to have much of a track record. It remains to be seen whether computerized algorithms like those required to hail a ride or find a bed for the night are sufficient to successfully match physicians to job opportunities.

First-time locums docs might appreciate the guidance and hand-holding offered by traditional agencies. Old-timers who know the ropes are better suited for computerized agencies. As of this writing, none of these new agencies are NALTO members.

On Your Own

You may work locum tenens with or without an agency. I've done it success-fully both ways.

One of my best assignments stemmed from a conversation with a col-league who offered me a temporary position at the academic medical center where he was chief of neurology. Because I was dealing with a good friend and a famous medical center, I felt comfortable negotiating a contract regarding housing, reimbursements and salary. Had there been problems with our negotiations, there was no third party to intervene.

In this case, the academic center had a full staff to assist with credentialing and licensing, as well as travel and lodging. This assignment turned out to be very pleasant and successful for a lot of reasons, but the informal arrangement was decidedly unusual.

Outside the Box

Spontaneous, successful assignments like the one above, which lasted a year and a half, are one of the delights of working locum tenens. The intentional impermanence and elasticity of employment opens up the prospect for unique experiences.

Getting Started

You can sign up with a locum tenens company without financial investment or commitment. NALTO requires member companies to obtain consent before submitting your resume to any of their clients.

Your Resume

Keep it brief, only one or two pages. Essential details should appear on the first page. Include the following:

- Permanent and locum tenens employment history
- Month and year of all locum tenens jobs
- Areas of specialization
- All states where you are licensed
- Leave abstracts, journal articles, and poster presentations for the end-these academic achievements hold little interest for locum tenens employers

Don't bother with nonclinical employment, family profile, or hobbies. Employers don't care if you own a restaurant, have three kids, a pet chinchilla, or climbed Mt. Kilimanjaro. On the other hand, if you used to work as a hospital chief medical officer, that might be worth mentioning.

Locum Tenens Credentialing

If the agency is going to accurately represent you to a potential employer, it will need more than your resume. Be prepared for paperwork as the locum tenens company initiates its own credentialing procedure. During this process, endeavor to establish an amicable relationship with a specific agent who will soon try to get you the best assignments.

If there are dark chapters in your past, it won't pay to keep them a secret. Painful though it may be, come clean regarding any history of malpractice or criminal activity. Significant mental health or substance abuse issues may also need to be revealed. If it's unclear what you need to disclose, consult with an attorney. Incomplete, inaccurate or misleading responses will likely backfire. False statements discovered on an application may be sufficient cause for termination.

In addition to your agency application, keep your ear to the ground for local opportunities. A temporary position may arise in your own community due to a medical group expansion, illness, maternity leave or other reasons. A local assignment is a low stress way to dip your toe into the locum tenens world. If you already have a full-time job, make sure the locums opportunity doesn't breach a non-compete clause or otherwise conflict with your current contract.

Some locum tenens agencies, like CompHealth, also offer permanent placement services. Staffcare provides permanent placement through sister companies Merritt Hawkins and Kendall & Davis. If you are considering a permanent job, it may be advantageous to select an agency that recruits for both locums and permanent positions.

Recommendations

If you are interested in giving locum tenens a try, browse a few agency websites and check out opportunities. Locum tenens companies often host booths at national medical meetings, which is a great place to get acquainted. Once you are ready, select at least one NALTO-endorsed company. Find an agent who understands your needs. Sign up for a brief assignment and see what happens.

*At first blush, it may seem incongruous that the author of a locum tenens book considered a permanent job. However, it's perfectly rational as the line between locum tenens and permanent positions is thin indeed. Current permanent job contracts are often as brief as a year, shorter than some locums assignments (Wilner 2016). I view permanent jobs as "extended locum positions." A locums job may also serve as a "try out" for a permanent position, allowing both the physician and employer to test the relationship. This arrangement is known as "locums to perm."

CHAPTER 9

Hospital Privileges, State Licenses and DEA Registration

Credentialing

The necessity of credentialing for hospital privileges and state licenses is arguably the greatest obstacle to a stress-free locums life. Each new hospital assignment and state license requires an exhaustive (and exhausting) application process.

According to the Joint Commission, formerly the Joint Commission on the Accreditation of Healthcare Organizations (JCAHO):

> *Credentialing is the process of obtaining, verifying, and assessing the qualifications of a practitioner to provide care or services in or for a health care organization. Credentials are documented evidence of licensure, education, training, experience or other qualifications* (The Joint Commission Accreditation Ambulatory Care 2016).

State medical boards are renowned for the stacks of supporting documents they require as well as the inordinate length of time needed to grant a license. Hospital privileging works independently of state licensing, but is similarly demanding. To add to the pain, both states and hospitals require much of the same information, resulting in a wasteful and expensive duplication of effort.

Each state or hospital begins their request for documents with your college diploma! Information is not accepted in a standard resume or curriculum vitae format, obligating you to struggle to fill in poorly sized blanks and check "yes" or "no" boxes for hours on end. Malfunctioning PDF forms add to the misery.

Hospital privileges generally do not take as long to acquire as state licenses, but weeks to months are customary. In my experience, polite attempts to expedite either process are futile. State medical boards and hospitals are independent entities with little accountability to you. They blithely proceed on their own glacial timelines.

Public Safety

The arduous credentialing process intends to protect patients from unqualified caregivers who have magically perpetrated the fraud that they are licensed physicians, are otherwise incompetent, or committed crimes upon society so heinous that they are unfit for the medical profession. No sensible individual would contest this virtuous goal. However, the time-consuming process that has evolved results in serious unintended consequences. For example, needy patients go unattended for months while a well-qualified physician awaits yet another formal review. A well-planned locum tenens assignment or even a career may be derailed (Chapter 8).

Personally, I think my patients would be better served if the countless hours I've devoted to credentialing had been spent reviewing American Academy of Neurology treatment guidelines or studying UpToDate. I've yet to save a life while working on a credentialing application.

Privacy

The rigorous credentialing process makes it clear than any invasion of your privacy is justified in order to protect patient welfare (Chapter 10). Don't be surprised, the way I was, when a "voluntary" drug screen was required in order to obtain hospital privileges at a venerable medical center. Credentialing also includes a criminal background check. No hospital wants to be caught issuing a white coat when an orange jumpsuit would be more appropriate.

Hospital Privileges

The Joint Commission is a nonprofit organization that accredits and certifies more than 20,000 health care organizations. It maintains a website (www.

jointcommission.org) that contains reams of rules and regulations sufficient to make any government agency proud.

The Joint Commission defines privileging:

> *...the process whereby the specific scope and content of the patient care services (that is clinical privileges) are authorized for a health care practitioner by a health care organization, based on an evaluation of the individual's credentials and performance (The Joint Commission Accreditation Ambulatory Care 2016).*

There are three other hospital accreditors, all with similar, but not identical, standards. Consequently, privileging requirements at one hospital may differ from another. Healthcare organizations may also implement requirements that go beyond the minimum, adding more unpredictability and complexity to the accreditation process (Matzka 2017).

One primary driver for this scrupulous credentialing regimen is the hospital's risk management department. According to one expert, "If the patient suffers an adverse outcome at the hospital, the hospital can be held separately liable for negligent credentialing if it is found that the credentialing was not performed appropriately" (Matzka 2017).

In addition, Centers for Medicaid and Medicare Services (CMS) requirements must be fulfilled by all health care organizations that provide services to Medicare and/or Medicaid patients.

For prospective members of the medical staff, minimal requirements include:

- Request for clinical privileges
- Evidence for current licensure
- Evidence of training and professional education
- Documented experience
- Supporting references of competence

CMS regulations require that hospital privileges delineate the scope of care a physician can provide based on character, competence, expertise, judgement, and training. The fact that subjective assessment of "character" constitutes a legitimate selection criterion allows hospitals enormous latitude whether or not to grant privileges.

The standards of all accreditors, including the Joint Commission, must meet CMS requirements. Not to be left out, states also impose hospital credentialing requirements.

Hospitals strive to follow Joint Commission as well as state and federal regulations to the letter. They have little choice. Loss of CMS credentialing would halt payments for Medicare and Medicaid patients, quickly hurling any public hospital and many others into bankruptcy.

Physician Requirements

Hospital credentialing requires evidence of board certification, Drug Enforcement Administration (DEA) and state medical licenses, malpractice insurance, medical school education as well as residency and fellowship training. Credentialing committees must also be informed of challenges to licensure or registration, voluntary or involuntary relinquishment of license or registration, voluntary or involuntary termination of medical staff membership, and voluntary or involuntary limitation, reduction or loss of clinical privileges. Letters from practitioners personally acquainted with you must testify to your "current clinical competence." You must also query the National Practitioner Data Bank (NPDB) to expose any adverse license, malpractice, or other malfeasance.

In addition, privileging may require details of past malpractice cases, immunization records (Chapter 11), explanation of work history gaps of as little as two weeks, continuing medical education (CME) records for the last two years and more.

You will likely need current basic life support (BLS) certification. Advanced cardiac life support (ACLS) or advanced trauma life support (ATLS) may be required for specialities such as anesthesia, critical care, ER, surgery and others. Check with the hospital medical affairs office when you apply.

Board Certification

Board certification used to be a "plus," a feather in your cap. Not only had you completed a residency in your chosen specialty, but you had mastered it.

Today, board certification (or eligibility) has become a de facto requirement for hospital privileges. One recent application stated, "…medical staff members not meeting the board certification requirement may lose hospital privileges, medical staff membership, and/or employment…" This hospital also required the dreaded "Maintenance of Certification" (MOC) for medical staff members.

Board certification or eligibility is usually a prerequisite to practice international locums in Australia or New Zealand. Canada and Singapore may also require board certification or an equivalent.

If you are not board certified, you are in the minority. Depending upon age group, 69-88% of physicians have attained American Board of Medical Specialities (ABMS) certification (Young et al. 2015).

Are there physicians out there perfectly qualified to practice medicine and not board certified? You bet. Current hospital privilege credentialing rules exclude them and represent "requirement creep."

If you have recently completed residency, you have a grace period of a few years to pass the boards. If you have been in practice for a while, it may be difficult to circumvent this requirement. Hospitals desperate for help may declare an exception. But with so many physicians working locum tenens to choose from, administrators would rather stick to their rules and hire a board-certified physician.

If you are not board certified, be up front with your staffing agent and hospital human resources department. First, the facts will emerge eventually, and better sooner than later. Second, your agent may know of facilities that impose less stringent requirements. These may not be the most desirable assignments, but they may prove better than none at all.

If you still have a number of years on your practice horizon, there's always the possibility of setting aside some study time, a few thousand dollars, and taking a shot at the Boards. It's time consuming and expensive, but will definitely improve your chances of getting more plum assignments. You may even learn something that helps your patients.

Bureaucracy Run Amok

Even if you would rather work in a clinic than a hospital, hospital credentialing may still be necessary. A few years ago, I treated patients in a local neurology outpatient clinic, with no hospital affiliation, duties or night call. My practice had *nothing* to do with the neighboring hospital.

However, one insurance company in the clinic's network would not reimburse for services unless all clinic physicians had hospital privileges. Since health insurance companies work in a parallel universe where common sense rarely finds footing, the clinic was unable to counter this requirement. Apparently, the insurer wanted the hospital to "vet" my credentials, saving them the trouble.

Securing hospital privileges became an odyssey worthy of Homer. After I had completed the lengthy application, the hospital was bought by a larger hospital system. This seemingly innocuous and not unprecedented event resulted in the following extraordinary correspondence (redacted for privacy):

> *2014*
> *Re: hospital credentialing*
> *Dr. Wilner,*
>
> *Good morning. I forwarded you a text with Dr. X's contact number in it, I hope you got that. I emailed the administrator at the hospital to see where your credentialing is and she has said it is dead, per her email below. I am upset with her as we had no notice from her on this and I forwarded your last responses and still had no feedback from her.*
>
> *So, unfortunately, another application has to be redone. She has informed us that their application has changed and the whole thing has to be resubmitted, not just the pages that may have changed.*
>
> *I do apologize for this inconvenience. I wish there was another way around this.*
>
> *If you have questions you can call and talk to Dr. X or myself. So sorry about this.*

You can imagine my reaction! In order to work at the clinic, I had to complete the new application...

Uncompensated Time

I recently completed the initial credentialing packet for a new hospital assignment (more information requests would follow). The application consumed 20 hours over two weeks and two weekends. Mind you, I have completed credentialing forms many, many times and keep all the required information at my fingertips. Still, somehow there is always a request that entails a lengthy search for an obscure document. One hospital that I worked at years ago no longer exists, which is true, sadly, for a number of my mentors. Good luck getting primary source verification from them (see below)!

Primary Source Verification

In order to receive Joint Commission or other accreditation, hospitals must perform "primary source verification" of your education, work experience, and other professional activities. Excessive concern about fraud invalidates submission of photocopies of your college diploma, medical license and residency certificates.

Primary source verification may be required for every school and training program you ever attended, every clinic or hospital where you ever worked, and every current or past state medical license you ever held (see below). While these verification measures may seem reasonable at first blush, my own experience reveals just how absurd these requirements really are for one who already has a state license. For example, is it reasonable to request medical school transcripts from a licensed physician in practice more than 30 years? Is it conceivable that I successfully practiced medicine all this time without ever attending medical school?

For each requested document, you must pay the required fee, then follow up to make sure the necessary material gets sent and arrives at the medical board. The more documents requested, the more likely one will get "lost in the mail."

Such concerns are not hypothetical. Requested documents gone missing delayed my Tennessee medical license for an additional two months.

Thankfully, primary source verification is not required for everything on the colossal credentialing checklist. For example, American Medical Association Physician Masterfile Profiles can be used to verify U.S. or

Puerto Rican medical school graduation and training programs approved by the Accreditation Council for Graduate Medical Education (ACGME). Many hospitals accept these.

According to the Joint Commission Comprehensive Accreditation Manual for Hospitals (CAMH) medical staff guidelines (MS.06.01.01):

> *Determining the competency of practitioners to provide high quality, safe patient care is one of the most important and difficult decisions an organization must make...The development and maintenance of a credible process to determine competency requires not only diligent data collection and evaluation, but also the actions by both the governing body and organized medical staff.*

The onerous standard of primary source verification bogs down credentialing. The process requires multiple contacts with third parties who have little to gain by providing the requested information. Primary source verification adds substantial expense as most organizations require a fee to forward necessary documents.

You will be the one to request information from schools, training programs, and places of employment. The responsibility that all this paperwork arrives at the hospital medical affairs office or state medical board also falls to you. This obligation puts you in the middle, calling to make sure the precious information was sent, then contacting the hospital or state medical board to make sure it was received. As a mere physician, obviously you are not trustworthy enough to convey these sensitive documents directly.

If you have worked locum tenens for a while, you may have practiced in many hospitals. To provide documentation of all past work experiences for credentialing, it helps to keep track of the dates and contact information of your supervisors.

Primary source verification is perhaps prudent to perform once, but not repeatedly. For example, even though my epilepsy fellowship at the Montreal Neurological Institute has been confirmed by every hospital I have ever worked at during the last 30 years, I will have to request yet another verification letter before my next locum tenens assignment.

Imagine how many hours it must take to provide and validate primary source verification for all the schools, training programs, and hospitals that

I've traversed during my medical career. Multiply that number by the nearly one million physicians in the U.S.A. to arrive at the millions if not billions of dollars in fees and lost productivity devoured by primary source verification. This seems a high price to pay to identify that one doctor in a million, or maybe two, who misrepresented their credentials.

If you studied overseas, fulfilling the requirements of primary source verification may be no small chore. If you trained in some unfortunate war-torn country or your medical school is defunct, I feel your pain.

People who support primary source verification must be the same ones in the airport security line who nod approvingly while government employees diligently frisk your wheelchair-bound elderly mother. Never mind that you and Mom miss your flight!

Daily Verifications

Primary source verification may not suffice in these days of litigation paranoia. Since license status may change due to expiration or suspension, computer programs exist for hospitals to perform daily primary source verification (www.evercheck.com). The existence of such products, as well as companies whose sole mission is primary source verification, testifies to just how concerned administrators are that one of their physicians should somehow be caught practicing without a license.

Peer Recommendations

One of the beauties of locum tenens is the flexible schedule it offers. But be careful not to extend your nonclinical time for more than two years. Hospital privileging committees require peer recommendations testifying to your clinical expertise within the last two years, an impossible requirement if you haven't done any clinical work in that timeframe. Volunteer work doesn't count.

Similarly, state licensing may require recent clinical work. The state of Texas, for example, requires one year of full-time work within the last two years prior to your application. Florida has a similar requirement, insisting on two years of clinical work within the last four.

These blanket regulations that limit state licensing and prevent qualified physicians from practicing are ill-conceived and detrimental to patients. Such laws also needlessly obstruct qualified retired physicians who wish to return to clinical practice.

The understandable concern is that physicians without recent hands-on patient contact will forget their skills. While much is made of daily advances in medicine, the process of history taking, physical examination, laboratory review, counseling and follow-up, which makes up the vast majority of a practicing physician's day, undergoes little change year to year.

The requirement for continuous or near continuous clinical work may be more applicable to surgical specialties. Skill levels vary among surgeons, and it is generally appreciated that those with high volumes tend to achieve better results. Surgeons must also keep up with technological advances, such as laparoscopic techniques, which have dramatically changed traditional routine procedures such as cholecystectomy and hysterectomy.

While it is a maxim that "practice makes perfect," it is unclear how much practice is necessary to maintain surgical skills (Merrill 2016). This probably varies from surgeon to surgeon. Cumulative, as opposed to recent experience, probably also counts. Current regulations ignore variability of individual talent and skill. As a result, surgeons who have been absent from their specialty for relatively brief periods of time find it impossible to return (personal communication, Leonard Glass, MD, Physician Reentry and Retraining Program, 2018).

It's conceivable that if a physician closed the door on medicine in order to golf intensively for a couple of years, he or she could fall so far behind as to become professionally irretrievable. While it is normal for anyone coming back from a long vacation to feel a bit rusty their first day back, the presumption that dedicated and experienced physicians will become incompetent after a year or two away from clinical practice is preposterous.

Make Friends

A piece of excellent advice comes from hospital privileging expert, Kathy Matzka, CPMSM, CPCS:

> The medical services professional (MSP) at the hospital is the main
> point of contact for the provider wishing to get privileges. In many

hospitals, this is a one-person office and the MSP is performing all medical staff support functions in addition to credentialing. It is very difficult to keep one's head above water. Locum tenens are usually needed ASAP, meaning the MSP has to drop what he/she is doing and work on the application. Very frustrating. If the LT wants to score some points and possibly get his/her application processed faster, he/she should take a few minutes to make an introductory phone call to the medical staff services department and connect with the person who will be processing the application. Ask if there is anything he/she can do to help. This will make the applicant seem like a person, rather than just another application that needs to get processed (personal communication, 2017).

Never Ending

The hospital privilege process never ends. Once bestowed, privileges must be renewed at least every two years. Re-credentialing requires additional information such as peer review and professional practice evaluations. Some hospitals may not award privileges to locum tenens physicians for the full two years, but only for the duration of the assignment. Many hospitals will grant privileges to locum tenens physicians, but not medical staff appointment, an acknowledgment of the physician's temporary status. Consequently, locum tenens physicians have little if any voice regarding improving hospital policies and procedures.

State Licenses

Itinerant locum tenens physicians often seek assignments outside their home states. Because no federal licensing system applies to the majority of U.S. physicians,* practicing medicine in another state requires an additional state license. State licensing usually takes 3-6 months, which can be a very long time in the fast-moving world of locum tenens opportunities (Chapter 8).

The possession of multiple state licenses optimizes the chances of securing a locum tenens opportunity in the desired time frame. One locums agency recommends that physicians carry a quiver of at least five licenses. One busy physician I spoke to had accumulated 14 licenses.

Over the years, I collected 13 state licenses. I let one expire this year because it was for a telemedicine assignment that won't likely recur (Chapter 8). I've let three others drop, as I don't anticipate needing them again. Nine may still be too many, as they are a challenge to keep up with and expensive to renew. Since I fantasize about retiring to a beach someday, I plan on keeping my Florida license.

A spreadsheet with license start and expiration dates will come in handy (Chapter 15). Before I relied on my trusty spreadsheet, my Massachusetts license accidentally expired. In my defense, I had recently moved and never received the renewal notice. When I realized it two days after the fact, I immediately telephoned the Massachusetts Medical Board. Just getting someone to answer the phone was a monumental achievement. When I finally succeeded in speaking to a person, the agent was unsympathetic. Rather than let me renew the license over the phone and accept a check by FedEx, and maybe a late fee, the agent insisted my license had "expired." The only solution was to start from scratch and complete a new license application. When I asked why the board didn't email or call to inform me of the pending expiration, the spokesperson retorted that they were much too busy to provide personalized service!

The new Massachusetts medical license application required several full days of tedious secretarial work as well as a hundreds of dollars. It's unclear how the board's bureaucratic decision benefited patient safety.

Application Abyss

The credentialing process for state medical licenses echoes the same, sorry story as credentialing for hospital privileges. Applications include endless requests for malpractice information, notarized attestations, recent passport-sized photographs, professional references, training history, work experience and my all-time favorite, a trip to the police station for fingerprinting and a criminal background check.

Simply navigating medical board websites to locate the application can prove challenging. State health departments issue licenses for a vast array of professionals, including acupuncturists and massage therapists. Trying to locate the proper physician application can become a perverse version of "Where's Waldo?" There are even "applications within applications," such as a "Declaration of Citizenship," required by the state of Tennessee.

State medical boards tend to be understaffed, difficult to reach by phone and slow to respond to email requests. One state board took *more than a month* to respond to a simple application question by email. Were state boards a business, rather than government agencies, they would have long since passed into bankruptcy.

Which States?

Think about where you want to work and why. Have you always wanted to explore Alaska's wildlife, Colorado's ski slopes, or Florida's sandy beaches? Do you dream of moving to Hollywood to launch your screenwriting career? Are you considering the Pacific Northwest as a lifelong home? Has the sunny Southwest tempted you as the ultimate retirement destination?

Ask your staffing agent which states offer assignments in your speciality. For example, neurology assignments in my home state of Rhode Island are few and far between. But North and South Dakota frequently offer positions. Consequently, I obtained a South Dakota license and worked in Sioux Falls with great success. It's a bit off the beaten track, but people were friendly, the hospital up-to-date, and pay scale generous. If you are interested in steady locum tenens work, your agent can recommend the states most likely to keep you busy.

A license in hand puts you at the front of the line. It's good for the employer, because they need you now. It's good for the agency, as they won't have to shepherd you through the tortuous licensing process and pay the fees. It's good for you, too. You won't have to complete another application and add another license to your spreadsheet. You're ready to work and get paid.

As one staffing agent observed,

> *The physicians with several state licenses work with us the most. It's that simple. If you have multiple licenses, we can present you in multiple states, meaning more job opportunities, which means more money for you.*

On the other hand, maintaining multiple medical licenses will tax your administrative skills (Chapter 15). Renewals occur on different dates and cost hundreds of dollars each. CME requirements differ for each state and must be tracked (Chapter 17). State sponsored surveys, updates and other announcements add clutter to your inbox.

Application Materials

The Federation of State Medical Boards (FSMB) provides a list of requirements on their website (www.fsmb.org). Your state license application may include all of the following:

- Summary and verification of medical education and residency training.
- Summary and verification of other medical licenses, current and past.
- Documentation of licensing examinations (FLEX, NBME, USMLE).
- Dates and location of every hospital where you held privileges.
- Work history with explanation of "gaps" of more than 30 days when you were not employed.
- At least two current references.
- Documentation of specialty board examinations.
- Documentation of two prior years of continuing medical education (CME).
- Federal Bureau of Investigation (FBI) criminal background check.
- Fingerprint cards.
- Summary of malpractice actions.
- Citizenship declaration.
- Jurisprudence examination.
- Attestation to physical and mental fitness.
- Release of liability for any harm you suffer due to the myriad of unnamed administrators prying into your private life.

Secretarial Skills

If the above list seems like a boatload of administrative work, you are correct. You must hone your secretarial skills. Clerical mistakes on application forms can result in weeks of confusion and delay.

I once checked "no" instead of "yes" after a long list of "no" responses was interrupted by a poorly worded question that required a "yes," something like, "Are you not encumbered by substance or alcohol abuse, mental illness, or any other incapacitating illness?" My bad for checking "no," but gee whiz! The application had to be resubmitted with the correct answer, slowing the approval process by more than a month.

Staffing agencies will help complete necessary forms. Of course, you are responsible for their responses, so it may be easier to do the work yourself. You may be the only one who knows the names, dates, and addresses pertaining to your training and work history. Answers to personal health questions can only be done by you, not the staffing agency. And only you can trot down to the police station, try and avoid eye contact with the ne'er-do-wells behind bars, and get fingerprinted for your criminal background check.

Medical Jurisprudence

A few states, such as Mississippi, Nevada, Oklahoma and Texas, require a medical jurisprudence examination. For those of you who regret not going to law school, here is your chance to revel in the arcane rules and regulations that govern medical practice. I've completed a couple of these examinations. Had I nothing else to do, I might have enjoyed this foray into the unfamiliar territory of medical regulations.

For my Mississippi jurisprudence exam, the state board kindly provided a 100-page study guide that covered a wide array of legal topics. Some rules were self-evident, such as the duty to obtain a license, complete CME, and release medical records to patients upon request. As might be expected, numerous pages were devoted to controlled substance prescriptions.

Some sections of the Mississippi administrative code were more personal. Could you guess it's a practitioner's duty to promptly report to the Board of Medical Licensure if they become seropositive for hepatitis B, C or HIV? For neurologists and physical medicine specialists, there are rules regarding delegation of electromyography (EMG) procedures. The booklet also contained instructions on how to properly terminate a patient-physician relationship, should that become necessary.

Luckily, much of the material was common sense and didn't require excessive study. For example, "Attempting to obtain, or renewing a license to practice acupuncture by bribery or misinterpretation," was subject to disciplinary action. No kidding.

It is possible that jurisprudence examinations impart some useful information. I leave it to you to decide whether this benefit justifies yet one more hoop for physicians to jump through in order to obtain a medical license and get to work.

Active Practice

As described above for Florida and Texas, you may be required to demonstrate that you have recently been in active clinical practice in order to obtain a medical license. If you are a physician who spent the last few years writing health care bills like Senator Bill Cassidy of Louisiana, worked in a pharmaceutical company, kept your eyes glued to a microscope in a basic science lab, or were otherwise employed in the health care industry without patient contact, you will still be ineligible for a medical license in these states.

For Florida and Texas, the only way to obtain a medical license is with recent medical experience, and the only way to obtain recent medical experience is with a medical license. Joseph Heller would have loved this Catch-22!

Once obtained, medical licenses do not require recent clinical experience for renewal. CME credits and a few hundred dollars will suffice.

Hospital privileging committees require recommendations based on recent clinical experience, often within the last two years, for all applicants. Consequently, physicians who interrupt their medical careers to climb Mount Everest, raise a few children, run for office, sail around the world, write a book, or pursue other nonclinical adventures with the intention of rejoining the workforce may find it impossible to regain hospital privileges *even though they have maintained an active state license*. (This fact alone is a great argument for practicing locum tenens at least once a year while engaged in nonclinical pursuits.)

Timeline

Obtaining a medical license takes weeks to months. Six months is not unusual. Here is an actual response from a state medical board, received after my latest license application:

We will work with your application in the order it was received.
You will receive a deficiency letter from our office. Please let us
know if we can be of further assistance to you.

Warm regards,

Observe the foregone conclusion that no matter how hard I worked on the application, no matter how many hours I dedicated to transcribe every requested detail of my professional life, so many requirements exist that no application could possibly escape a "deficiency letter." As an example of the operating speed of the medical board, I didn't receive any further contact for several months.

A slow medical board response can lead to painful ramifications. For one assignment, by the time the state license was issued, the job was gone (Chapter 8). No work, no pay. The hours spent in the application process? Wasted.

Hospital privilege committees are just as pokey, typically meeting only once a month. They won't even start the process until you have a state license. Since they may require additional information after their first meeting, it can take a month, two, or even longer to obtain hospital privileges. One hospital that needed me right away actually delayed my start date because they couldn't complete their own work in a timely fashion! Even a physician happily employed for years without incident at a different hospital across the street will undergo the same, protracted privileging process.

License Verifications

If you already possess one or more medical licenses, you will discover that it is more of a hindrance than a help when attempting to obtain an additional one. Each license must be verified as part of your new state license application. Verifications reveal disciplinary actions or suspensions and attest whether the licenses are in good standing. You must arrange for these license verifications to be sent directly to the state board considering your application. Even if you are licensed in a state that allows online license verification, which anyone with a computer could do, you still have to make a formal request, pay the fee, and follow up on the delivery.

An online company, Veridoc (www.veridoc.com), streamlines this state license verification process. I use them regularly. Unfortunately, only certain states accept Veridoc. The other states must be dealt with by postal mail.

Interstate Medical Licensure Compact

Why hasn't there been an outcry regarding the unwieldy burden of state licensing? Probably because 79% of physicians hold only a single state license (Young et al. 2014). Once they complete the onerous licensing process and have it behind them, there's no reason for them to agonize over it.

However, the percentage of physicians with more than one license is increasing because of the growth of locum tenens and telemedicine opportunities. Both of these factors increase the need for practitioners to acquire multiple state medical licenses.

Telemedicine has been recognized as a cost-effective delivery system, particularly in the rural Midwest. This economic realization has fueled enthusiasm for the development of a federal medical license that would be valid in all 50 states.

In the meantime, telemedicine physicians must obtain a state license for each state in which they serve patients. Depending upon the geographic scope of practice, multiple state licenses may be necessary.

Despite what you may have heard, there is no reciprocity between state medical boards. The new Interstate Medical Licensure Compact (IMLC) is a baby step in that direction. The IMLC is the desperate response of state medical boards to the threat of an efficient, portable, single federal medical license. Such a license would minimize the importance of state medical boards as well as their budgets, if not do away with them altogether.

IMLC does promise to streamline the state license procedure, but still requires physicians to obtain individual state licenses and comply with each state's requirements and fees. In addition, IMLC requires an additional nonrefundable $700 fee. (Bureaucrats have to eat, too). The IMLC began issuing "letters of qualification" in April, 2017.

Here is the justification of the IMLC from the horse's mouth, the Laws of Minnesota, 2015, Chapter 55-S.F. No. 253:

> *In order to strengthen access to health care, and in recognition of the advances in the delivery of health care, the member*

states of the Interstate Medical Licensure Compact have allied in common purpose to develop a comprehensive process that complements the existing licensing and regulatory authority of state medical boards, and provides a streamlined process that allows physicians to become licensed in multiple states, thereby enhancing the portability of medical license and ensuring the safety of patients. The compact creates another pathway for licensure and does not otherwise change a state's existing Medical Practice Act.

In the tradition of bureaucracies worldwide, rather than simplify the current licensing process, the IMLC created an additional, costly secondary system. But not every physician is eligible. Participation requires that a physician comply with the following:

- Graduate from an accredited medical or osteopathic school.
- Pass the USMLE or Comprehensive Osteopathic Medical Licensing Examination (COM-LEX-USA) within three attempts.
- Successfully complete graduate medical education.
- Hold specialty certification by the American Board of Medical Specialties or the American Osteopathic Association's Bureau of Osteopathic Specialists.
- Possess a full and unrestricted medical license by a member state.
- Never been convicted, received adjudication, deferred adjudication, received community supervision, or deferred disposition for any offense by a court of appropriate jurisdiction.
- Never held a license authorizing the practice of medicine subjected to discipline by a licensing agency in any state, federal, or foreign jurisdiction, excluding any action related to nonpayment of fees related to a license.
- Has never had a controlled substance license or permit suspended or revoked by a state or the United States Drug Enforcement Administration.
- Is not under active investigation by a licensing agency or law enforcement authority in any state, federal, or foreign jurisdiction.

Physicians who meet these and other criteria are eligible for reciprocity regarding their "static qualifications" from their home state to the new state. This eliminates the need for primary verification (see above) of medical education, training, and other matters of record.

The IMLC should speed speed things up. As of November 2, 2018, the IMLC included 24 states. Texas and Florida, as you might imagine, have not joined. Up-to-date IMLC information is available at: www.IMLCC.org.

Logic is Optional

There is no obvious logic underlying the arduous process of state licensing. For example, a Nevada Medical Board representative proudly boasted that her state had one of the most demanding license applications in the country. She actually congratulated me for surmounting the Herculean administrative hurdles. Her attitude seemed oddly incongruous (and perhaps partly responsible?) with the fact that Nevada residents suffer a severe doctor shortage!

License Cost

Similarly, logic fails to apply to the cost of a medical license. Why does the Alabama Medical Board charge a paltry $75 while the great state of Mississippi requires $600?

Nevada may be getting the message. As of July 1, 2018, their registration fee dropped from $750 to $350. But they still maintain a $600 application fee and $75 for a criminal background check, a total of $1,050.00.

Medical license costs are exorbitant. Let's use a license fee of $500 as an example. Multiply the nearly 1,000,000 U.S. physicians times $500, and you realize that the price of health care includes approximately an extra half-billion dollars every year, simply for physician license fees.

One could argue that state medical boards do very important work that needs to be supported. That argument might contain a few grains of truth, but not a half-billion dollars worth.

Knowledge Examination

Patient safety is always the catchphrase proposed to defend onerous administrative requirements. But if states were genuinely concerned about patient safety, why don't they insist on a recent medical proficiency exam? Passing test scores from the Federation Licensing Examination (FLEX), National Board of Medical Examiners (NBME), or United States Medical Licensing Examination (USMLE) are sufficient, no matter when they were taken.

Sometime in the latter part of the 20ᵗʰ century, I passed Part III of the National Boards. It was the last medical test I've ever needed for any of my 13 state medical licenses. *Not one state* requires an additional test of medical knowledge or skills no matter how long it's been since you passed the boards.

The convoluted process of state licensing tests patience and fortitude, not medical knowledge. Please don't misunderstand. I'm not recommending yet another examination. Just making a point about arbitrary and burdensome license requirements...

Helpful Services

Two services are available (for a fee) to assist in forwarding your verified credentials for hospital privileging or state licensing. The best known is the Federation Credentials Verification Service (FCVS). (I rely on FCVS regularly.) Another one is EPIC (Electronic Portfolio of International Credentials), not to be confused with the electronic medical record software of the same name. EPIC addresses credentialing needs of foreign medical graduates and is hosted by the Educational Committee for Foreign Medical Graduates (ECFMG).

Expiration Dates

Once you've initiated your state license application, it may take months for all the supporting documentation you've requested to arrive. While the board will notify you once the application is complete, they may not tell you what's missing until you ask.

You may understandably be reluctant to pester the state board, but you must monitor your paperwork's progress. Applications that remain incomplete for six months may expire. Absurd as it may seem, this has oc-curred. It almost happened to me. Here's the short version:

Several months into my application for a Tennessee medical license, the board informed me that they were missing two state license verifications, one from Minnesota and one from North Carolina. If all the required information was not received within two weeks, the board would terminate my application.

This terse email arrived on the background of prior reassurance that my application was "complete" and awaiting final review. Suddenly, the application was not complete and headed for the shredder! The thought that months of ad-ministrative drudgery, countless letters and emails, not to mention hundreds of dollars of fees were futile caused me to drop what I was doing and telephone the medical board. When I finally made contact, they reassured me the application wouldn't be cancelled. The notification was just a "form letter." Whew!

In fact, both state verifications in question should have already been neatly tucked into my application. I knew this because of my spreadsheet (Chapter 15). The Minnesota verification had been requested from Veridoc, which had sent five other verifications. Since Veridoc sent them all together, *on the same printed page*, how could only Minnesota be missing? Where could it be? I called Veridoc and explained the situation. They were kind enough to resend it without charge.

Since the state of North Carolina doesn't work with Veridoc, I had to contact them to resend their verification. Never mind that I have not had a North Carolina license for more than 20 years! What value could this ver-ification bring to my application? Nonetheless, according to the Tennessee medical board, without this crucial document, no action could be taken on my already delayed state license application.

The above anecdote is testimony to the Kafkaesque medical license application process. It illustrates the vigilance and perseverance applicants must exert to succeed. If all documents don't arrive by a certain date, the medical board may cancel your application. You will have to begin again from the beginning, all for the privilege of going to work!

A New Assignment

You definitely have to think twice before accepting an assignment in a new hospital or state because of the credentialing burden. Even though your staffing agency will assist and pick up much of the cost, you will still end up doing considerable secretarial work.

If you land a job in a new state, give yourself at least six months to get licensed. If the license comes through sooner, the employer may take you right away. If not, a vacation might fit in nicely while waiting for the new job to start.

Once you get some locums experience under your belt, the same job may come up again. I've worked at a wonderful medical center in Minnesota twice in a period of two years (Chapter 8). The second time around, although my hospital privileges had expired, they were easily renewed because the primary source verifications had previously been completed. I already had a state medical license, DEA and CSP numbers, knew the facility and the people. I hope to return for a third round someday.

No New License Needed

If all of the above sounds like too much trouble, special situations exist where you can use the license you already have for locums assignments, no matter which state they are in. The Department of Defense (DOD), Indian Health Service (IHS), and Veterans Administration (VA) employ civilian locum tenens physicians and accept any valid state license. Government jobs also tend to have lower volume and less demanding call requirements. Providing care to active duty military and culturally diverse communities offers additional personal and professional rewards. Correctional institutions also often need physicians.

If you are leaning towards long-term assignments, these government options may be particularly attractive. The VA requires at least a three-month commitment, while the DOD expects six months. While you don't need an additional state license, each of these organizations employs their own credentialing system that includes a thorough background check. I'm working on a VA application now, and so far it's not too bad. I've been told there's more paperwork to follow, but it's not happening fast...

DEA Registration

I always assumed that one Drug Enforcement Agency (DEA) registration number sufficed because it was issued by the federal government. *I was wrong.*

Each state requires a unique DEA number if you prescribe controlled substances (Chapter 4). Ignorance of this fact has caused at least one physician to lose his medical license.

For example, if you practice in more than one state (i.e, locums on the weekends in Massachusetts and weekdays in your Rhode Island office), you will need two DEA numbers. If you only practice in one state at a time, you can switch your DEA number to the relevant state. For example, I worked in Arizona for two years. When I completed that assignment to begin work in Minnesota, I simply switched the DEA number from Arizona to Minnesota online. It was not necessary to apply (and purchase) a new one.

Technically, physicians are allowed to utilize the DEA registration of the hospital where they work in lieu of obtaining their own, if the hospital agrees. That's not something I have ever done.

In order to prescribe controlled substances, some states require a state narcotics certificate termed a Controlled Substance for Practitioners (CSP) or Controlled Substance Act (CSA) in addition to your DEA registration. This is completely redundant.

You will also have to create an online account for the state's Prescription Monitoring Program, also known as a prescription monitoring and reporting system (PMRS). This program requires you to check a patient's medication history prior to writing a new controlled substance prescription.

There is no federal PMP. If you practice in multiple states, you must register for PMPs in each state. If you treat patients who come from adjacent states, your state PMP will not pick up their out-of-state controlled substance prescriptions.

Make sure you have the proper DEA registration, CSP certificate and PMP for each state where you work. The onus is on you to get it right.

Unintended Consequences

Hospital accreditors like the Joint Commission and CMS set high standards for physician credentialing in order to protect patients from unqualified physicians. While well-intentioned, these cumbersome requirements delay hospital privileges, discourage applicants, and impede qualified physicians from caring for needy patients. Locum tenens physicians, who may work for relatively short periods of time in diverse locations, suffer disproportionately from these man-made practice barriers.

For me, the administrative boondoggle of hospital privileges and licensing is the biggest downside to *The Locum Life*. Flexibility become fantasy when bureaucracy becomes a barricade.

If you focus on a few geographic areas rather than take a scattergun approach to state licensing, paperwork should be manageable. An experienced staffing agent can recommend the states most likely to offer a steady stream of assignments (Chapter 8).

Recommendations

Plan your work schedule with the knowledge that hospital and state credentialing rivals the gestation time of large mammals. Follow up promptly on requested documents and references. An application that remains incomplete too long may expire.

If you enjoy an assignment, express your enthusiasm to everyone involved. If they invite you back, you will already possess hospital privileges, the proper state license, and be merrily on your way to another fulfilling and profitable locum tenens experience.

*Physicians practicing in government institutions such as correctional facilities, Indian Reservations, military, and the Veterans Administration need only a single state license.

CHAPTER 10

Physician Privacy

Privacy Invasion

Inquiries regarding your education, malpractice, medical training, mental and physical health, work history and even criminal record appear on employment, hospital privilege, medical license and other applications. Many of these questions constitute a brash invasion of your privacy (Wilner 2015). Some are insulting.

Nonetheless, you must answer honestly. Your job and medical license are at stake. Armies of clerks earn their paychecks by fact-checking your responses. Someone somewhere probably gets a bonus if they catch you in a lie.

HIPAA

What about HIPAA (Health Insurance Portability and Accountability Act of 1996)? Doesn't that federal law protect privacy?

HIPAA protects the sanctity of the patient-physician relationship and personal *medical* information. HIPAA doesn't do squat for a physician's personal privacy.

Criminal Background Check

I submitted to a criminal background check when applying for a neurohospitalist position in Minneapolis, MN. Minnesota law requires the Department of Human Services (DHS) to conduct criminal background checks on individuals who come in direct contact with patients. The background check includes your address, birthdate, criminal convictions and reports of maltreatment of minors, vulnerable adults and more. Participation in a police lineup isn't required (yet).

You might expect that your valuable personal and professional information would be limited to individuals with a "need to know." Indeed it is. However, that need to know list includes:

- Agencies with Criminal Record Information in other States
- County Agencies
- County Attorneys
- County Sheriffs
- Courts
- Federal Bureau of Investigation
- Health-Related Licensing Boards
- Juvenile Courts
- Local Police
- Minnesota Department of Corrections
- Minnesota Department of Health
- Non-Licensed Personal Care Provider Organizations
- Office of the Attorney General

In addition to this extensive network that may share your personal information, my hospital privilege application included this disturbing paragraph in fine print.

> I authorize the Entity and its Agents to consult with any third party who may have information bearing on my professional qualifications, credentials, clinical competence, character, mental condition, physical condition, alcohol or chemical dependency diagnosis and treatment, ethics, behavior, or any other matter reasonably having a bearing on my qualifications for participation and authorize such third parties to release such information to the Entity and its Agents.

In other words, your signature grants authorization to anyone, including your aerobics instructor, spouse, ex-spouse, ex-spouse's sister, third grade teacher, or recently fired secretary to weigh in on your application. The right to incriminate you extends to nearly everyone on the planet.

The notice that spells out these unabashed privacy invasions reassuringly states, "Individuals who do not have disqualifying characteristics will not be disqualified." Feel better now?

Drug Testing

Hospitals routinely require urine drug screens prior to awarding physicians the "privilege" of treating patients (Lowes 2013). At an academic hospital in Arizona, a clerk politely informed me that the drug screen was "voluntary" and instructed me to sign a form acknowledging the test's voluntary nature. Then I was handed a plastic cup. It was not stated in so many words, but no drug screen, no job.

This bold perversion of the definition of "voluntary" is worthy of Orwell's 1984's "Newspeak." Welcome to the brave new world of medicine. Yet more urine was analyzed for my most recent neurology faculty position in Memphis, TN.

Although I don't use illegal drugs, my concern was that an early morning poppy seed muffin might trigger a false positive test, government investigation, and the end of a spotless record. Next thing you know, I'm unemployed and on the run like Richard Kimble from *The Fugitive*. Far-fetched? It's happened more than once (Upton 2014, McMullin 2017).

Of course, no patient wants a drug-addicted doctor, but the fact that screening tests are far from perfect constitutes a legitimate argument against mandatory "voluntary" drug testing. False positives occur as often as 10% of the time (Laino 2010).

Note to self: Switch to blueberry muffins on drug testing days.

The Burden is on You

When applying for hospital privileges, the burden is on you to supply the hospital with the necessary information to allow them to decide whether you qualify.

From a Nevada hospital:

> *I understand and agree that I, as an applicant for medical staff membership, have the burden of producing adequate information for proper evaluation of my professional competency, character, ethics, and other qualifications and for resolving any doubt about such qualifications, during such time as this application is being processed, I agree to update the application should there be any change in the information provided which may affect the application or its outcome.*

Financial and Personal Interests

Hospital privilege applications may require disclosure of your financial and personal conflicts of interests as well as those of your family. Here are required disclosures from my Nevada hospital application:

- Consulting, equipment or space lease, employment, licensing, royalty, services arrangement or other financial relationship.
- Interest that contributes more than 5% to a member's annual income or the annual income of a family member.
- Position as a director, trustee, managing partner, officer or key employee, whether paid or unpaid.
- Ownership through sole proprietorships, stock, stock options, partnership or limited partnership shares, and limited liability company memberships.

Fingerprints

Most, but not all, states, require fingerprints for a criminal background check prior to approving a medical license. Fingerprinting is a time-consuming process, and it doesn't always succeed. Experts have told me that doctors' fingers are notoriously difficult to print. More than one joked that this occupational anomaly might prove advantageous for another profession.

Fingerprints for my South Dakota license had to be taken three times to satisfy Federal Bureau of Investigation quality standards. Each attempt required a trip to the local police station and a half-day off from work. As the licensing process dragged on, precious weeks evaporated when I could have cared for patients and earned a paycheck.

At the police station, I remember a strange feeling of dread as a detective led me through a succession of corridors followed by steep stairs to the basement. As we progressed through the maze, a series of sturdy locked doors yielded to his oversize skeleton key. Finally, we reached the fingerprinting station in a dim, airless alcove flanked by a row of jail cells.

Much to my dismay, the unique arches, loops and whorls on my fingertips resisted printing. Repeated chemical baths, inking, and xeroxing failed to produce acceptable fingerprints. The large clock on the wall ticked away unpaid hours.

While the detective and I wrestled with the fingerprint machine, several hung-over gentlemen monitored our efforts with faint amusement from their holding cells. Every now and then one of them took a break from the show to retch in a bucket.

Finally, at my suggestion, the frustrated officer resorted to typing a letter. Addressed, "To Whom It May Concern," the missive stated that the indistinct prints he obtained from my fingers were as good as anyone could get.

My fingerprinting experience wasted the officer's time as well as my own. Since I don't have a criminal record, the painstaking process didn't make anyone any safer.

Peer Recommendations

Hospital privileges require peer recommendations regarding current clinical competence (Healthcare Facilities Accreditation Program). They don't need to be "big" names, just physicians who worked with you and have good records themselves. Choose them well.

If you are not certain whether colleagues will write favorable recommendations, provide an easy out. Ask whether they have sufficient time to complete the task, since you know they are so busy. If they hem and haw, gallantly offer to find someone else. You'll do yourself a favor and your colleagues will breathe a sigh of relief. Most physicians would prefer to decline than write an unfavorable letter.

Malpractice

Locum tenens agencies always ask about malpractice (Chapters 11 and 12). If you are lucky enough to avoid malpractice claims and other professional complaints during your career, a clean record will catapult your application to the top of the pile, all other things being equal.

If you are over 55 years old, it's better than 50/50 that you have had a malpractice claim. Although malpractice suits are common, most claims are dropped. Of those that go to trial, physicians win 90% (Kane 2010).

Nonetheless, whether or not you were at fault, you must report any pending, current or past malpractice suits. Ask your attorney to prepare a concise written statement. Staffing agents don't need a long explanation, just a response.

Skeletons in the Closet

I keep a human skull near my desk, but no skeletons in the closet. Not everyone is so lucky. Your hospital privilege and license applications contain approximately 20 questions asking whether you have abused drugs, committed crimes, had licenses suspended, practiced without liability insurance, or engaged in other unsavory behavior.

Some requests overreach reasonable bounds. Here are two examples from my Minnesota license application:

- Have you ever been a defendant in any malpractice lawsuits, had any malpractice settlement, or have any pending? (The simple fact of being a defendant in a case that was ultimately dismissed isn't anyone's business.)
- Have there ever been any criminal charges filed against you?...This also includes any offenses which have been expunged, or otherwise removed from your record by executive pardon. (Again, just because a case was filed doesn't mean it bears repeating. And doesn't expunged mean the case has been erased?)

If you do have a criminal record (you knew those Hell's Angels days would catch up with you), you must answer "yes" and briefly explain.

Inaccurate information or "alternative truths" will invalidate your application and could get you fired if the true story ever comes to light.

Take your time and carefully read the questions. Don't check the wrong box (like I did) (Chapter 9). As always, when it comes to legal matters, consult an attorney.

Do these invasions of privacy constitute necessary due diligence to "protect the public" from nefarious physicians? Perhaps. But there may be few doctors left in practice when only saints can pass the vetting.

Release of Liability

Not only must you provide all requested information, you must also sign a "release of liability." Should any personal fallout occur because of the medical staff's privacy invasion, as long as it occurred, "in good faith and without malice," you have no legal recourse.

Application-Job Disconnect

Fortunately, the disagreeable nature of hospital privilege and license applications bears little relation to job quality. In fact, one of my best assignments, the one in Minneapolis, had the most degrading application. Disgusted and exhausted, one night I nearly tossed it in the trash.

Upon arrival at the hospital, I learned that the callous questions did not correlate with the pleasant reception, congenial colleagues, stimulating cases, and grateful patients. It turns out that most of the people who review these documents are hapless victims of the beastly bureaucratic process, just like us.

A Grand Inquisition

Whether intended for a locum tenens or permanent position, job applications subject physicians to unprecedented invasions of privacy. Hyperbolic concerns for patient safety consistently trump a physician's right to privacy.

The process becomes more burdensome with locums because it repeats more frequently. Each new application triggers a grand inquisition that encompasses your education, training, malpractice history, National Data Bank Reporting, substance use, even your mental health.

If your responses reveal a blemish or two, don't worry. Perfect doctors are few and far between. If you are otherwise well qualified, you will still land the assignment.

Privacy in the Workplace

If you manage to successfully acquire hospital privileges, a state license, and actually start working, congratulations! But the invasion of privacy doesn't stop there.

If you need to access your personal email or text messages while at work, do it with your smartphone or laptop. Do *not* use one of the ubiquitous hospital computer workstations. In addition, avoid using your handy work email for personal communications.

Why? Somewhere in your institution's mountain of official documents will be a clause something like this:

> *No Expectation of Privacy: All Systems, including hardware and software are property of the Organization and may be recovered at any time without prior notice. Files, documents, emails, photos, and etc. are not the private property of any employee, and users should not maintain any expectation of privacy with the respect to any usage of the Organizational Electronic Systems. All personnel waive any right to privacy while using Organizational owned Information Systems (Policy documents of a hospital in Memphis, TN).*

Wow!

My Minnesota hospital's "confidentiality agreement" expressed a similar policy, "All of my [hospital name] accounts are subject to auditing and/ or monitoring."

Make no mistake. Management, whoever they are, owns your emails and texts. Big Brother, anyone?

Gossip

Whether you like it or not, your arrival at a new hospital or clinic will attract attention. While you may not be subject to continuous video surveillance, you might as well be. At least in the beginning, curious eyes and ears belonging to staff and colleagues will monitor your every move.

Hospital cafeterias, coffee shops and corridors are among the world's greatest gossip centers. Your skills, manners, even appearance will become hot topics of conversation. It's not for nothing that *General Hospital* is the longest running television soap opera.

Remember, you work in a fishbowl. The only place you might find privacy is behind the restroom door. But don't count on it.

Recommendations

Answer application questions honestly and briefly. Budget time for this professional interrogation. Don't take it personally. At work, don't expect a shred of privacy. Keep your personal life personal.

CHAPTER 11

Malpractice-Avoiding Litigation

Introduction

Malpractice suits provide patients an avenue for financial redress in case of harmful medical errors. This legal recourse aims to protect patients from careless treatment and compensate those who are injured. Nonetheless, many feel that our adversarial legal system negatively affects routine patient-physician relationships.

Malpractice suits have become an unfortunate fact of life, affecting 50% of practicing physicians. While injured patients receive compensation from successful malpractice claims, litigation costs tend to exceed patient compensation, benefiting the "system" more than the patient (Studdert et al. 2016). Malpractice cases also tend to drag on, prolonging the agony of the patient's injury, delaying compensation needed for medical bills and other financial obligations, and casting a black cloud over the defendant physician's head. The average time from injury to resolution is five years (Studdert et al. 2016).

Medical malpractice suits tend to occur without warning, and the financial and emotional cost can be devastating (Sklar 2017). Even physicians who have not been sued suffer the constant, looming threat of a malpractice suit. Chronic worry as well as the practice of nonproductive and expensive "defensive medicine" often results. As with any disease, prevention is the best medicine.

Most physicians carry malpractice insurance, (professional liability insurance), although some choose to go without. Cost depends upon your insurance provider, malpractice history, specialty, state and other factors.

Malpractice insurance covers the investigation of a claim, defense, and settlement (up to the policy limits).

One of your first priorities before actually seeing patients is to decide whether you want professional liability insurance. If so, you must select the appropriate policy and figure out who is going to pay for it. Locum tenens companies provide professional liability coverage. Their largesse is not altruistic. If you do have a legal problem, the company doesn't want to get dragged into it because you don't have insurance.

No one wants to get sued. Let's see what you can do to avoid a malpractice claim in your locum tenens practice.

What is Malpractice?

A physician who provides quality care has not committed malpractice just because a bad outcome occurs. In order to establish medical malpractice, three legal criteria must be fulfilled.*

1. Patient-physician relationship
2. Substandard care
3. Significant harm from substandard care

The first category is usually, but not always, straightforward. For example, if your neighbor asks for medical advice for his cousin in Arkansas, and the cousin dies following your advice, it is highly unlikely you will be susceptible to a medical malpractice suit because a patient-physician relationship had not been established.

However, had your neighbor brought his cousin to your home a few times, and you opined about his condition, there might actually be a case. Accordingly, risk management attorneys urge caution regarding "curbside" consults.

To satisfy the second criterion, substandard care, the plaintiff's attorney must demonstrate that the quality of care was below what would be expected in that medical community. To complete the argument for malpractice, the attorney must demonstrate that the substandard care resulted in injury.

Whether medical malpractice has occurred is a legal matter and not always straightforward. The difference between quality and substandard care may not be clear cut. If the care was substandard, it still must be shown it led to the untoward outcome.

Malpractice and Locum Tenens

Because malpractice suits are so common, a past claim won't blacklist you from locum tenens work if the rest of your application looks good. Several malpractice claims would, however, become problematic.

With respect to the probability of malpractice while working locum tenens, there is no inherent increased risk (Cash 2017). However, the fact that you will frequently work in an unfamiliar environment requires additional vigilance on your part to avoid errors. In addition, don't forget to obtain appropriate malpractice insurance (Chapter 12).

One exception: if you work in military, Indian Health Service, Veterans Administration or certain other federal health care facilities, you will not need your own professional liability insurance. In these environments, the federal government acts as insurer.

Many physicians work locum tenens to supplement their salaries from regular jobs. Even though they already have professional liability insurance, an additional policy is necessary to cover the locum tenens work. Major staffing agencies such as CompHealth and Staffcare routinely provide malpractice coverage.

Risk Stratification

Many circumstances determine whether or not a malpractice suit occurs. For example, certain medical specialities are more susceptible. In an American Medical Association review of 5,825 physicians, nearly 70% of general surgeons and obstetrician gynecologists were sued compared to less than 30% of pediatricians and psychiatrists (Kane 2010).

Another variable is the number of patients seen. The more years in practice, the greater the likelihood of a lawsuit. More than 60% of physicians 55 and older had a malpractice claim.

Based on the above information, one strategy to avoid malpractice claims would be to pick a low risk speciality and retire early. While this conclusion stems logically from the data, it is not a practical path for most physicians.

Pathophysiology of Malpractice Suits

In order of frequency, the six most common reasons for malpractice suits are (Kreimer 2013):

1. Clinical judgement error (missed or delayed diagnosis)
2. Technical error (woops!)
3. Communication error
4. Patient behavior
5. System failure
6. Inadequate documentation

A quick look at the above list reveals that #6 is the easiest one to eliminate. There's no excuse for poor documentation. You don't need to be a genius to document your work, and the electronic medical record (EMR) makes it easier. Typing into the EMR should be quick. If you can't type, consider learning. EMRs are not going away.

If you can't learn to type, ask whether the EMR supports dictation. Many hospitals utilize "Dragon Dictation," which allows one to dictate directly into the medical record as fast as you can talk (Chapter 16). Such automated transcription won't be perfect, particularly if you have a strong accent, but it will be a whole lot better than nothing.

The hard part of documentation is to budget the time to explain what you did and why. This task becomes easier when you realize how important it is. With proper documentation, the medical record attests to your thoughtful and excellent medical care. Any extra time required is more than rewarded by staying out of court.

Your next target is #3, communication error. This one is a little tougher, but communicating with patients may be at least as important as what you actually do to them. I have often had patients thank me for explaining their care and prognosis, even though I had just delivered very bad news. Grateful patients are unlikely to file malpractice suits.

Patients may be dissatisfied with your communication skills because you are rushed, use medical jargon, appear preoccupied, look at your pager, phone or watch too often, speak with a difficult to understand accent, or don't seem to care. A poor outcome may magnify deficiencies in your bedside manner in the eyes of the patient and his or her family.

Limit interruptions as best you can during patient and family conferences. Daily updates need not take long. Patients are already overwhelmed with their illness and treatments. I have found that "short and sweet" explanations are far more effective than meandering discussions encumbered with hypotheticals. The simple phrase, "Let's see how it goes and reassess tomorrow," communicates caring, continuity, and allows one to move on. I use it a lot.

The need for effective communication applies not only to patients and their families, but to colleagues and staff as well. For example, nurses are more likely to carry out orders accurately (and with more enthusiasm) if they understand what you want and why.

Ask a trusted colleague to assess your verbal and nonverbal communication skills. If you need work, get it. (It's a rare physician who could not stand improvement in this department.)

The remaining four causes of malpractice are tougher to prevent. All of us make clinical judgement and technical errors from time to time. Obviously, mistakes need to be minimized.

Part of a physician's job is to stay up-to-date, consult other physicians when in doubt, and always be the best doctor you can be. Additional strategies are to avoid sleep deprivation,** take the necessary time for decision making and procedures, use order sets and follow established protocols (all easier said than done, I know!). Consider risk management continuing medical education (CME) to reinforce these practices.

Patient behavior is pretty much beyond your control. There will always be "difficult patients," many of whom have personality disorders. Patients who adhere poorly to prescribed therapy are more likely to experience poor outcomes. Some will blame you.

When you treat a difficult patient, recognize the problem and summon every ounce of diplomacy you can muster. You can politely disagree when a patient insists that herbal supplements cure cancer, but don't argue. Share these special patients generously with your colleagues.

All institutions have system errors that can lead to poor patient outcomes. They can only be repaired by motivated staff and administrators. Here's an example of one system error that was actually fixed.

Years ago, one of my partners suspected that his patient might have a brain tumor, but the head CT report was normal. Turned out an addendum did indeed report a tumor. However, the addendum was typed on an additional page, which had been misfiled. Weeks later, while reviewing the chart, he came upon the errant report. When my partner finally caught his breath, he called the patient. This glaring oversight could have become a major lawsuit.

One institution where I worked successfully addressed this "addendum" problem. All radiology addendums are now posted at the *beginning* of the report, not the end. Voila! A simple solution that costs nothing to implement. Makes sense, but someone had to recognize the problem and take administrative steps to implement the fix.

If you want to take an active role on hospital committees to improve efficiency and reduce medical errors, God bless you. However, because of your temporary status as a locum tenens physician, you probably won't have the opportunity. Even if you did, it's doubtful you will stay at the facility long enough to effect meaningful change.

Social Media

Sharing intimate details of our lives with the world is all the rage these days. But participation in social media represents a potential risk for physicians entrusted with confidential patient information. The Health Insurance Portability and Accountability Act (HIPAA) empowers the federal government to exact heavy fines for disclosure of protected health information (PHI), whether willful or accidental.

If you enjoy Facebook, Instagram, Snapchat, Twitter, or other social media, restrict their use to your private life. Resist the urge to share your exciting patient care experiences.

One local ER physician lost her job due to an innocent Facebook post about an interesting patient involved in a motor vehicle accident. Even though she was careful to keep the patient's personal details anonymous, the hospital argued that because it was a small town, some readers could deduce the patient's identity because the crash had been reported in the local news. As soon as the doctor realized there was a problem, she removed the Facebook post.

Despite the inadvertent nature of her error, the hospital board found her "guilty of unprofessional conduct." She was reprimanded, fined, and fired. You have nothing to gain and everything to lose by sharing PHI on social media.

Be Nice

Physicians, like other human beings, make mistakes. In order to properly defend you, your malpractice company will ask for a heads-up regarding any adverse events that may culminate in a malpractice claim.

Medical mistakes don't necessarily evolve into malpractice suits. Most patients injured by a medical error do not sue (Studdert et al. 2006). In particular, if patients like their doctor, they will not seek a malpractice attorney (Kreimer 2013).

Kindness and compassion go a long way to making patients feel better during their illness and may also keep you out of court. For example, my partner, who has an excellent bedside manner, was never sued in the brain tumor snafu described above.

My advice is the same as your mother's. Be nice to everyone (whether they deserve it or not). This includes administrators, whom you may someday need on your side. Always practice the art of medicine by taking time to listen and heal.

Application Forms

Hospital credentialing and state medical licensing applications will inquire about your malpractice history. If you have to check "yes" regarding a malpractice claim, you already have an attorney. Ask for a succinct written response to respond to these obligatory questions.

The Good News

It goes without saying that a malpractice suit is an enormous headache. However, only 5% of medical malpractice suits go to trial. The rest are dismissed or settled. Of the 5% that are tried in court, physicians prevail in 90%.

Malpractice Avoidance Checklist

The first step in avoiding malpractice in your locum tenens career is to comply with federal, hospital and state requirements. Make sure all of the following are in order:

- Current medical license in the state where you will practice. This may take months to obtain.
- DEA number linked to the state.
- "Controlled license registration" (CSP). Some states require them, some don't.
- Enrollment in a "prescription monitoring program" (PMP). If you practice in multiple states, register for PMPs in each state.
- Current Basic Life Support (BLS).
- Hospital privileges correspond to responsibilities and tasks. For example, if you are not privileged to give "conscious sedation," don't do it. You won't have a leg to stand on if something goes wrong. On the other hand, you are always "privileged" to act appropriately in an emergency, so don't hesitate to start CPR even if your certification card has expired.
- Medical malpractice insurance (Chapter 12).
- Up-to-date continuing medical education that includes specific courses for your state (Chapter 17).
- If you practice telemedicine, you need licenses for all states where your patients are located as well as your home state.
- Clarify on-call requirements. Do you need to be available 24/7? What is the required response time if you are paged? Are you obligated to come to the hospital when called?
- Clarify coverage responsibilities for other physicians, physician assistants and nurse practitioners. Will you directly supervise anyone or be responsible for practitioners off-site? Will you have to cosign their notes? Will you be asked to write or renew prescriptions for patients you have never seen? Such additional responsibilities are not unusual. Locum tenens physicians often work in institutions that are short-staffed, which is why you were hired in the first place.

- Will you field telephone calls from hospital-affiliated clinics, ERs, and patients after hours? If so, must you document these calls?
- Do you know how to use the hospital/clinic electronic medical record (Chapter 16)? If not, better learn fast (Wilner 2013).

Recommendations

Practice quality medicine, document thoroughly, be nice to everyone, and get a good night's sleep. Make sure you have malpractice insurance.

*This is a practicing neurologist's view of malpractice. For the real deal, consult your attorney.

**The most serious medical error I ever made was after a sleepless on-call night during my internal medicine residency. Luckily, the patient survived my error. I refuse locum tenens opportunities that require prolonged sleep deprivation.

CHAPTER 12

Malpractice Insurance-Tail or No Tail?

Nobody's That Perfect

Nearly all physicians protect themselves from the potential financial devastation of a malpractice suit with professional liability insurance. Even physicians so perfect they can't conceive of making a mistake need coverage because they may be blamed for a bad outcome that is not their fault (see below). The cost of attorney fees, expert witnesses, and associated expenses can be considerable, even if the physician is ultimately vindicated.

Going Bare?

A minority of physicians choose to forego professional liability insurance due to its cost and/or the belief that their uninsured status makes them less appetizing targets for hungry malpractice attorneys (Weger 2017). Some states allow this practice, but most require minimum levels of professional liability insurance.

Just as in the practice of medicine, there is never a "one size fits all" answer when it comes to the business of medicine. In my research, I could not find any insurance expert who advised "going bare." I'm not recommending it here. There are many resources such as your attorney, staffing agent, state medical society, and peers who can advise on the wisdom of acquiring professional liability insurance.

Choosing an Insurer

When purchasing a professional liability policy, there are many factors to consider. These include the financial strength and reputation of the insurance company, policy terms and conditions, premium, definition of a claim, "consent to settle" terms, and tail coverage (see below).

Nuisance Suits

Even if you don't do anything wrong, you may be subject to a malpractice suit. Such events do not create fond memories. An ER physician related this story:

> The nuisance suit involved a woman who had knee replacement surgery. She saw her orthopod on Friday, saw me on Sunday with a superficial redness at the incision site. I drew a CBC, sed rate, and prescribed antibiotics (this was before the "everything is MRSA" days), and tried to contact her physician, albeit unsuccessfully. She saw her orthopod on Tuesday, who aspirated the area. On Thursday, she was called back and admitted for a "washout." I wanted to fight it because the standard of care was not to aspirate a superficial infection. The insurance company settled for "nuisance" value-$45,000, after spending money to bring it to trial. After the jury was selected, but before the case started, and against my wishes of settling, they settled. The plaintiff's witness for EM care was a "rent a whore."

Agency Perk

Major locum tenens agencies such as CompHealth and Staffcare provide malpractice insurance. When I've worked without an agency, the hospital purchased the insurance.

Dollar Limits

Standard professional liability insurance limits for locum tenens physicians are $1,000,000 per claim and $3,000,000 aggregate per provider. The "aggregate" is the most that will be paid out during a policy term. Some states, such as New York and Virginia, require higher caps. Know the limits for your state.

The policy "face sheet" lists the insurance company name, dollar limits, and coverage duration. Scan and save the face sheet (Chapter 15). Your next job will require proof of prior insurance.

Occurrence or Claims-Made?

If you have decided, like most physicians, to obtain professional liability insurance, your next step is to purchase the appropriate policy. Two categories of policies exist, "claims made," and "occurrence." It's crucial to know the difference because they offer different degrees of protection.

The statute of limitations for filing a claim of medical malpractice varies by state, ranging from one year in Kentucky to ten years in Missouri. What happens if you complete an assignment in Missouri with no problems, but a malpractice suit surfaces five years later? Are you covered? If you have an occurrence policy, rest easy. If you have a claims made policy, expect sleepless nights.

The terminology of "occurrence" and "claims made" makes about as much sense as the words partial and complex when applied to seizures. Their meaning is not self-evident. Even people in the insurance industry sometimes confuse them.

Claims made policies expire when the policy terminates, which likely coincides with the last day of your assignment. Occurrence policies last more or less forever. As you might imagine, claims made policies are less expensive than occurrence policies because of the shorter duration of protection. To obtain continual protection, a claims made policy needs to be supplemented by an "Extended Reporting Endorsement," colloquially called a "tail."

Claims made policies, not surprisingly, are offered more commonly than occurrence policies. Occurrence policies are not even available in every state.

Tail or No Tail?

The practical solution to the limited duration of claims made professional liability insurance is to purchase a "tail." This tail covers your work indefinitely.

If your agent or hospital offers claims made insurance, ask who is responsible for purchasing the tail. It may be you, and it may be expensive. The price may be up to five times the cost of the original policy, payable as a lump sum. When I worked with Staffcare recently, they only offered claims made coverage, but did pay the tail. CompHealth, another major locum tenens agency, also offers claims made malpractice with tail coverage.

An alternative to a tail is "retroactive coverage," which also provides long-term protection. Retroactive coverage may be difficult to obtain if you move out of state.

My new staff neurologist position has the potential to become permanent. My employer only offered claims made insurance. If I leave within two years, the tail cost is mine. If I stay longer, my share progressively decreases to zero at five years. I guess they want me to stay...

I wasn't happy with this arrangement, as five years seems like an eternity to a seasoned locum tenens physician. My employer wouldn't negotiate this point, and I didn't want to risk the job to argue it (Chapter 13). Even if I last only three years, the cost should be manageable.

Scope of Coverage

Professional liability coverage applies only to medical negligence. Examples include missed diagnosis, incorrect medication administration, or a forgotten sponge in the abdomen.

Professional liability insurance does not cover other unfortunate situations such as alteration of records, criminal acts, misrepresentations on applications or sexual improprieties with patients or staff.

I remember the sad case of one of my colleagues. He was a distinguished neurologist in his seventies accused of trading narcotics for sex during after-hours house calls. The state medical board responded to the complaint and launched an investigation. In addition, the patient filed a civil suit.

Although constantly bemoaning the charges, he never denied them. The local paper enthusiastically published daily updates of the sordid story.

It was an awkward time for his fellow neurologists as we wanted to be supportive, but didn't know whether he was guilty or not.

To add to his misfortune, my besmirched colleague discovered his malpractice insurance did not apply. Attorney fees and related expenses were his responsibility.

What happened? He abruptly retired. Since he no longer possessed an active medical license, the medical board lost purview and dropped the investigation.

However, the civil suit proceeded. The court found him guilty of sexual impropriety, but determined the patient was equally complicit. The doctor was fined one dollar!

The neurologist's poor judgement ruined a fine medical career and reputation. Due to the constant press coverage, the case also cast a pallor on all the local physicians. The only upside, if you could call it that, was my busy practice got busier when he retired.

Don't Forget Your Umbrella

An umbrella policy may be purchased in addition to a primary professional liability policy. An umbrella kicks in when the limits of the primary policy are exceeded. For example, if your primary policy pays $1,000,000, and the settlement is $2,000,000, a $1,000,000 umbrella policy covers the difference. Of course, this assumes that all the umbrella policy requirements are satisfied. Discuss with your staffing and insurance agents whether an umbrella policy is appropriate for your situation.

Recommendations

Obtain a copy of your professional liability policy from your staffing agency or hospital. Check that dates and limits are correct. If you have a claims made policy, make sure it comes with a tail. Consider an umbrella. If you have questions, ask an insurance expert.

CHAPTER 13

Contracts and Compensation

Need for Speed

Today's medical marketplace offers abundant opportunities for qualified locum tenens physicians. Given the law of supply and demand, the hospital or clinic likely needs you more than you need them. You possess far more negotiating clout as a locums applicant than as a potential permanent employee.

Locum tenens is all about delivering clinical service as soon as possible. Both you and the locum tenens company are on the same page when it comes to the need for speed.

Staffing agents will strive to get you to work as soon as possible for several reasons. First, hospitals may use several locum companies simultaneously, with the contract going to the first one that serves up an acceptable candidate. Second, the sooner you start, the sooner the agent earns a commission. Third, they want to make you happy so you'll work with them again.

More than a Handshake

While the locums agency cannot control the maddening complexity of hospital privilege and state license applications (Chapter 9), they can design their own contracts. These tend to be brief and uncomplicated. Neither party benefits from a complicated document that requires time-consuming back and forth with expensive attorneys.

For example, my locum tenens contract with CompHealth consisted of a mere five pages. Conversely, when I recently signed on for a permanent job, the contract consisted of 14 pages followed by a four page "offer letter."

(Unlike my locum tenens contracts, this permanent one merited an attorney's review. See below.)

A one-page confirmation letter will accompany each locum tenens assignment. This document spells out essentials such as your schedule and pay rate. Your signature signals your commitment.

If you work without an agent, you will receive a contract from the hiring facility's human resources department. Of course, no contract substitutes for integrity, trust and effective communication between both parties.

Typically, locum tenens contracts allow either party to withdraw up to 30 days before the start date. After that, penalties kick in for cancellation.

Compensation

How much money you earn depends on how well you negotiate your rates and how much you work. Locum tenens physicians are generally paid hourly.

Compensation rates vary depending upon specialty. For example, neurosurgeons receive a higher hourly rate than neurologists (go figure!). In addition, compensation varies according to the urgency of the assignment, geographic location, type of facility (private vs. government), and other factors (Chapters 3 and 4). If you don't mind working on holidays, you can earn considerably more than the usual rate.

You can get a starting point of what your compensation should be by examining annual compensation surveys, such as the Medscape Physician Compensation Report (www.medscape.com). You may hear that locums physicians make more money than traditionally employed physicians. This is often true, but comes with a caveat-taxes and benefits are not included (see below).

If you are a 1099 employee (Chapter 14), taxes will not be withheld from your paycheck. Consequently, to equate your locum tenens hourly rate to traditional after-tax take-home pay, you have to factor in tax payments. In addition, since you will not receive traditional benefits such as disability, health or life insurance, these costs must also be subtracted from your salary.

Compensation for each assignment will vary. An assignment letter stipulates dates of service, hourly rate, overtime rate, holiday rate, on-call rates, and payment for returning to the hospital after hours. If you contract with a staffing agency like CompHealth or Staffcare, your paycheck will come directly from them, not the hospital or clinic where you work. Direct deposit is the industry standard and easy to arrange.

The details of the particular assignment dictate whether you have any leeway to negotiate higher pay. Some locum tenens positions are highly sought after with many applicants. In these situations, you have little leverage. However, if you are willing to work overnight ER shifts in Anchorage, AK, in January, you have a shot at a bigger paycheck.

I have found staffing agents eager to work with me to negotiate a pay raise (Chapter 8). Their efforts are not completely altruistic. If I'm happy and earning more, they're happy and earning more.

As might be expected, government assignments offer less salary flexibility. However, pay rates for the the military, Indian Health Service, Veterans Administration and correctional institutions may still be negotiable.

Check Specifics

I risk overstating the obvious, but it is necessary to read and understand the provider services agreement (contract) and assignment letter. If you have questions, ask your agent.

At the very least, these documents should spell out who will pay for hospital privilege and license expenses, housing, travel (air and rental car), and professional liability insurance (Chapters 11 and 12). You will be expected to keep your state medical license, DEA number, and CSP (if needed) current.

The contract will mandate that you maintain proper billing and patient care records, avoid illicit drug use, and generally stay out of trouble. Failure to uphold these basic requirements may result in "termination with cause," which you definitely want to avoid. Getting fired will seriously hinder attempts at future employment.

Do I Need an Attorney?

I have never hired an attorney to examine a locum tenens contract. First, these contracts are very straightforward (Chapter 4). Second, they are for temporary assignments that can be quickly terminated. For example, my CompHealth contract can be cancelled with a 30-day notice by either party for any reason. If an issue arises that can't be resolved, it can be easily remedied by leaving!

Nonetheless, if there's something in the contract that continues to perplex or trouble you, hire an attorney. A few hundred dollars spent up front on expert legal advice is a small investment when balanced against assignments that will yield thousands of dollars a week.

A Permanent Position

While working locum tenens, you may become interested in a permanent position. Your life circumstances may change (i.e., marriage, children, elder care) or you may stumble upon a work situation that is too good to leave. If so, consider a permanent contract.

A minority of locums assignments are "locums to perm," where the intention is to stay on as a permanent employee if both sides agree. Here is a recent example from Weatherby Healthcare:

> A North Carolina-based medical facility needs locum tenens neurology coverage. The provider will work 8-hour shifts on a 1:3 call rotation. The ideal candidate would be open to staying on as a permanent position. The locum tenens physician will see approximately 10-20 patients per day doing EMG, NCV, Botox, LP, stroke, and tPA procedures.

A contract for a permanent position is a horse of a different color. Early termination may trigger painful penalties such as loss of matching retirement benefits, the cost of an extended reporting endorsement (malpractice tail) (Chapters 11, 12), enforcement of a restrictive covenant, expensive relocation, and burned bridges that sabotage future employment. Although a typical locum tenens contract doesn't merit an attorney's review, a permanent contract does.

As a footnote, it's fair to say the word "permanent" doesn't mean what it used to when it comes to physician contracts (Wilner 2016). Many years ago, my first contract lasted five years, and I couldn't be fired without cause. Today, one or two year contracts are commonplace, and termination can occur by either party for any reason with 90 day notice. The Cleveland Clinic, for example, prides itself on one-year physician contracts.

Contract Changes are OK

Years ago, I took my first full-time academic position at a prestigious university. To say it was a "disaster" does not do justice to the experience, kind of like calling the sinking of the Titanic an "accident."

During initial contract negotiations, I asked to change the word "without" to "with." This three-letter change meant my position could only be terminated "with" cause, not "without." Otherwise, job security would have been at the administration's whim. The neurology department initially resisted, claiming the change constituted unnecessary paranoia on my part. In addition, it would require time-consuming legal review, resulting in a delay of my start date.

Since I wasn't in a hurry, I insisted. They made the changes.

My instincts had been correct. In the firestorm that followed, the precautionary edit supplied sufficient leverage to gracefully depart a dysfunctional department.

I learned one legal concept from this nasty encounter: the contract is king. When push comes to shove, it doesn't matter what was implied or promised, only what is written down. The lesson? If there is language that troubles you, discuss it with your attorney. If he or she agrees, insist on the changes. Once you sign, it's *your* contract.

Negotiating Tips

Most health care facilities plod along with rigid administrative structures that drag out contract negotiations. You may get lots of pushback like I did, e.g., "It will take the attorneys too long," or "The contract has to be the same for everyone."

If they really want you, your future employer can find a way around boilerplate language. Ignore lame responses and stand your ground.

No Hard Feelings

Don't be shy when negotiating a contract. The results are a lot more important to you than to the health care administrators on the other end. For you, this contract dictates details of your career, the culmination of years of education and sacrifice. For them, your contract is just another office chore.

Even if hard feelings develop, don't despair. I guarantee the administrator won't lose any sleep over it. Once you start work, it is unlikely you will regularly interact with anyone involved in contract discussions. Most hospital administrators take great pains to avoid messy patient care areas.

If the Job Doesn't Work Out

For a number of reasons, the reality of a locum tenens assignment may not match up with the rosy picture painted by a staffing agent. It may be the agent actually had little knowledge of the job. Or, in the style of many real estate agents, the staffing agent may have skewed the truth by excessively highlighting positives and minimizing negatives (Chapter 8).

Make sure your locum tenens contract has an escape clause. Maybe the patient load is overwhelming, the facility chaotic, or you've got an emergency at home. You need the freedom to give reasonable notice and leave if it's necessary for self-preservation while protecting your professional reputation.

If you're not happy at work, the facility may not be either. A sufficient volume of complaints empowers the facility to terminate you. If you get fired, that black mark will mar your resume forever. It's far preferable to leave on your own terms. An early departure may serve both parties well.

If it's a permanent job that doesn't work out, you've got some reorganizing and introspection to do. Maybe a good time for locums? It worked for me!

Recommendations

Read your contract. Make sure you understand it. Lots of locum tenens positions await your application. You've got leverage to negotiate. Consult an attorney if you consider a permanent position.

CHAPTER 14

Taxes and Business Expenses

Your Own Business

Congratulations! As a locum tenens physician, you now own a business. Small though it may be, this new enterprise consists of a cell phone, home office, laptop computer, stethoscope and most importantly, you.

You will probably work as an "independent contractor" rather than a salaried employee, although both situations are possible and can even co-exist (Chapter 1). From a tax perspective, you wear two hats; employer and employee.

Taxes

While this chapter hopefully offers some useful guidance, it would be a fool's errand to attempt to provide even the most general tax advice. To paraphrase Dr. Leonard McCoy, "I'm a doctor, not an accountant!"

U.S. tax laws constantly evolve. The abrupt, major changes in the federal tax code that occurred in 2018 have important implications for physicians. Self-employed locum tenens physicians take notice.

Although the Internal Revenue Service (IRS) posts many useful documents on its website, their relevance is not always clear. Interpretation of federal and state tax statutes requires expert assistance.

My certified public accountant (CPA) answers questions, estimates and files quarterly federal and state taxes, and has the last word on which business expenses are deductible (see below). I have used both a general CPA and a physician specialist CPA and have been much happier with the latter.

As a business owner, you may apply for a tax identification number to use in lieu of your social security number. A tax ID number does help separate personal and business finances and offers some degree of privacy for your social security number. I have a tax ID, but haven't found any great advantage to it. Ask your accountant if you need one.

The ability to contribute to a tax-deferred retirement account is a government perk for self-employed individuals and a topic in and of itself. Discuss retirement planning with your CPA and a financial consultant if you need one. Retirement planning is beyond the scope of this book.

Quarterly Taxes

Because the U.S. government doesn't trust self-employed individuals to save sufficient cash during the year for tax time, it requires quarterly payment of estimated taxes. These estimates are based on your prior year's tax return. Make discussion of quarterly taxes a high priority with your accountant, especially if this is your first year earning "extra income" as a locum tenens physician. There are penalties for underpaying.

Independent contractors do not have federal, Medicare, state or social security taxes withheld from their compensation. To safely budget your tax burden, assume that roughly 20-40% of your earnings will disappear as government tribute.

You will receive an IRS 1099-MISC form by January 31 of the following year that reports your locum tenens earnings. Every employer who paid you more than $600 in a 12-month period must issue this form. Your accountant tallies your gross income from these 1099s. When you add up all your paychecks from each employer, the amount should equal the 1099 reported income. Contact your employers if there's a discrepancy.

You may also work locum tenens as a salaried employee and your income will be reported on an IRS W-2 form. This arrangement is less common and typically for longer assignments. As a W-2 employee, taxes are withheld each pay period. You don't have to make quarterly tax payments on this income.

You must file tax returns in your state of residence as well as all other states where you work that levy a personal income tax. For example, if you live in Rhode Island, but take assignments in neighboring Massachusetts, you will pay taxes in both states. In order to avoid double taxation, your home state will typically credit the taxes paid in the other. (You don't have to worry about this if you work in Alaska, Florida, Nevada, South Dakota, Tennessee, Texas, Washington or Wyoming, as these states do not tax this type of income.)

In general, self-employed locum tenens physicians do not have state taxes withheld from their paychecks. However, California and Pennsylvania require staffing agencies to withhold a percentage of gross compensation from locum tenens physicians who work in either of these two states but live elsewhere.

During a calendar year, you might work in several states and have to file tax returns in each one. In 2017, I filed three state tax returns; Arizona, Minnesota, and Rhode Island. Multiple tax returns are another justification for a competent CPA.

Business Expenses

When it comes to determining whether an expense is tax-deductible, the IRS makes the rules, and a lot of them. As a self-employed individual, you are entitled to take certain business expenses as deductions against your taxable income. Your independent contractor related income and business expenses are reported on IRS Form 1040 Schedule C.

What's Deductible?

According to the IRS, business expenses are defined as "the cost of carrying on a trade or business." In order for a business expense to be deductible, it must be both "ordinary" and "necessary" for the business to make a profit.

For example, you might wish to purchase a Gulfstream G650 jet in order to perform house calls to Forbes 500 executives nationwide. Even if you could demonstrate that the jet was "necessary" for your unique business model, a Gulfstream G650 is not an "ordinary" physician expense and would not qualify.

On the other hand, as a locum tenens physician, you may pay renewal fees on five or more state licenses. As these state medical licenses are both "necessary" and "ordinary" for your work, they are legitimate business expenses.

It should be self-evident that any reimbursed expenses are not deductible.

For example, if you travel to Minneapolis every week to work at the county hospital, like I did, and the staffing agency reimburses the cost of your plane ticket, hotel and rental car, none of these are deductible. Because you were reimbursed, you suffered no "expense."

On the other hand, if you purchase a new stethoscope for your hospitalist work, and receive no reimbursement, that purchase is tax-deductible. A stethoscope is both "necessary" and "ordinary" for patient care.

Expenses must be documented. Save receipts and/or designate a single credit card for business use. Track business expenses on a spreadsheet (Chapter 15). For meals and incidental expenses, per diem rates may be more practical than saving each receipt. Per diem rates differ according to location and season. They are posted on the U.S. General Services Administration (GSA) website.

Personal, family, and living expenses are not business expenses. However, If you purchase a smartphone for both home and business use, the business use portion is deductible. Similarly, if you use part of your home as an office, a percentage of your home expenses (rent, insurance, repairs, utilities, etc.) is deductible.

Here is a list of typical business expenses for self-employed locum tenens physicians:

- Cell phone (business percentage)
- Computer
- Education (CME)
- Equipment
- Home office (business percentage)
- Internet (business percentage)
- Licenses
- Professional liability (malpractice) insurance
- Medical books, journals and dues
- Unreimbursed travel and lodging
- Uniforms (white coat, scrubs)
- 50% of meal costs during an assignment away from home

In addition to the above, I also take my Dropbox cloud subscription (Chapter 15) and Yahoo business mail as expenses. Netflix doesn't qualify. Ask your accountant which business expenses apply to your situation.

The IRS provides relevant information on its website (IRS.gov). For the full story, check out "Publication 535, Business Expenses."

Health Insurance

Independent contractors are responsible for their own health insurance. Typical policies may cost $1,000/month or more. In recognition of this financial burden, the IRS allows a 100% deduction for the cost of health insurance premiums. This important deduction has saved me thousands of dollars a year in taxes.

Do I Need to Incorporate?

A corporation separates business and personal finances. It shields personal assets from business liability and pays taxes differently than an individual. Creating and managing a corporation requires a lot of accounting paperwork.

One locum tenens physician I know hired an attorney to form a limited liability company (LLC). He wanted to cleanly separate the business from his personal expenses. He was also concerned he wouldn't have the discipline to save enough money for quarterly taxes.

The LLC set up only cost a few hundred dollars. His bookkeeper manages expenses and interacts with his accountant. He also contracted with a payroll company for his salary.

Personally, I haven't found it necessary to formalize locum tenens work via an LLC. For me, it would just add another layer of bureaucracy to an already overbureaucratized life. I try and focus most of my energy on earning income rather than managing it.

LLCs and corporate entities are valid business tools that may provide additional asset protection and other benefits. I've never needed them, but everyone's situation is different. Discuss with your CPA.

Recommendations

Hire a CPA experienced with physician clients and locum tenens. Allocate a big chunk of your paycheck for state and federal taxes. Don't forget to budget for expensive necessities like health insurance and retirement. Use a single credit card for business expenses.

CHAPTER 15

Mobile Office

The Peripatetic Physician

When you work locum tenens, you are a physician-in-motion. You will spend quite a bit of time traveling. The carefully organized file cabinets in your well-appointed home office will collect dust while you slave away on assignment in some distant locale. When I started working locums in the pre-internet days, loss of access to files was a frustrating problem. Not anymore.

Mission Control

The heart of your mobile office is your laptop. A high quality laptop is essential to run your business. A laptop is a critical tool that can facilitate your work or serve as a source of frustration (or both). Computer literacy has become as important to a modern physician, if not more so, than all the tools in a doctor's traditional black bag.

For example, my staffing agency requires that I complete a timesheet on their website. No more pen and paper. No more faxing (see below). The online system allows me to submit my hours and see when they have been approved. I can also set up direct deposit, create a W-9 form, submit expenses, and access helpful resources. All these tasks are easily accomplished with my laptop, an essential business tool.

Buy the best laptop you can afford. "Best" translates into the fastest processor, highest-capacity hard drive, lightest weight, longest battery life, most responsive keyboard, sharpest screen, and easiest to use software. Try and get as many of those features as you can. Since your laptop is a legitimate business expense, you can afford to splurge (Chapter 14).

It doesn't matter whether your laptop is a Mac or personal computer (PC). Apple computers use proprietary software and PCs mostly use Microsoft's Windows. As a long-time Apple user, I can safely say that any new Apple laptop will suffice. For PCs, numerous online reviews can help you choose. (The "Mac vs. PC" conundrum can't be resolved here.)

If your assignments take you off the beaten path, you may consider a "Toughbook." Two popular brands are Getac and Panasonic. These PCs are constructed to withstand physical abuse and are preferred by emergency medical technicians (EMTs), firemen and the military. Their ruggedness comes with a steep price.

Another option is a laptop that uses Google's Chrome operating system. Known as Chromebooks, they depend upon internet access and cloud storage. If you are a light computer user, these computers may be adequate and save you a bundle.

Both Mac and PC platforms operate software to accomplish any task a locum tenens physician might require. The choice of software is one of personal preference, but should take into account compatibility. For example, at the hospital, you will almost certainly use a PC. If you create files at home on your Apple laptop with Apple's productivity suite (Keynote, Numbers and Pages), they will not be viewable or editable on a hospital computer. This may be just as well, as it is preferable to keep your personal work private (Chapter 10).

If you must access your Apple files on a PC, you can use the Mac version of Microsoft Office (Excel, Powerpoint and Word). This software allows you to move files back and forth from Mac to PC without difficulty. Alternatively, use a PC at home, which will work more or less seamlessly with the PC at the hospital.

Tablets

State of the art tablets have become so powerful that they approach the functionality of laptops. Which one you use is a matter of preference. A device along the lines of a Microsoft Surface Pro or iPad Pro can serve as the hub of your traveling office.

If you know how to touch type, you will definitely want a keyboard attachment for your tablet, as typing is far easier than pecking on a screen. Modern tablets also have a stylus option. Tablets are also great for watching movies on the plane or after work in your hotel. Depending upon your budget, needs and preferences, you may end up with both a laptop and a tablet.

Software

To manage your own business, at the very least you should have a working knowledge of Microsoft Word for documents and Excel for spreadsheets (or their equivalents). If you give Grand Rounds or other lectures, you need Microsoft's PowerPoint, Apple's Keynote or similar presentation software.

If you are not fluent with these software programs, tutorials are available on YouTube and other websites such as Lynda.com. The "Dummies" series of books may be helpful. Your hospital may also offer classes, as basic software proficiency is necessary for nearly every employee.

Some physicians find computer software intuitive, others don't. At the risk of sounding politically incorrect, an inverse correlation between operator age and computer aptitude often exists. Given that three-quarters of locum tenens physicians are over 50, if you struggle with your computer, you are definitely not alone. None of us were born knowing how to copy and paste or calculate expenses on a spreadsheet.

Laptops and tablets are incredibly powerful tools, but their efficiency is user-dependent. Time devoted to learning how to use software to write a business letter, calculate expenses or create an engaging Grand Rounds presentation is a wise investment. While perhaps daunting at first, the more you use these standard programs, the easier they become. If you were smart enough to memorize the brachial plexus, you should have no problem.

Laptop Security

You must password protect your laptop or tablet to prevent random (or nefarious) people from accessing your personal information. A good password contains a complex assortment of letters, numbers, and symbols, but is not too complicated to remember. Each time you open your laptop, you must enter your password, so choose one easy to type. Consider a laptop with a fingerprint login, Apple watch activation, or other work around.

Cloud Storage

While your laptop's hard drive can accommodate thousands of files, the key to your mobile office's success is "cloud" storage. The cloud allows internet access to all your files via a password-protected account.

The cloud is merely someone else's computer or group of computers, really big ones in one or more secret warehouses. For example, Google's cloud service, Google Drive, stores data in Asia, Europe and the U.S.A. From your point of view, it doesn't matter where the files are as long as they are secure and you can use them 24/7. Just like email, you can access cloud storage files from a multitude of devices including computer, laptop, smartphone and tablet.

What you store in the cloud is up to you, but should include everything you need while on the go. For example, Controlled Substance for Practitioners (CSP) and DEA numbers, detailed work history, expense spreadsheets, medical licenses, password list, photo ID, resume, tax returns, travel reservations, vaccine documentation, the occasional downloaded movie and thousands of other files reside in my Dropbox cloud account. There's still room for more. Want to enjoy your photos during downtime? They can live in the cloud, every single one of them.

Crashproof

The merits of cloud storage include not only portability, but resilience. Unless a systemwide failure of Biblical proportions disables the cloud, your data are safely protected from fire, flood, theft and personal computer failure (see below). Cloud storage is not only more practical, but arguably safer than a hard drive backup (although you can do both, and I do).

Few of us have the discipline to regularly back up a hard drive and store it offsite to protect data from natural and manmade disasters. While that solution provides security, it is incredibly inconvenient. In addition, data protected in the offsite hard drive are inaccessible and unsynchronized.

According to PC magazine, the ten best cloud storage and file-sharing services of 2018 were: Apple iCloud Drive, Box, Dropbox, IDrive, CertainSafe, Digital Safety Deposit Box, Google Drive, Microsoft OneDrive, SpiderOakOne and SugarSync (Muchmore and Duffy 2018). The PC Magazine article details their pros and cons, which is probably more than you want to know. For the locum tenens physician, any of these would be suitable, and some would be overkill.

User friendly choices include Amazon Drive, Apple iCloud Drive, Box, Dropbox, Google Drive, and Microsoft OneDrive. These all offer free introductory accounts, which is a good way to get your feet wet. However, you will rapidly outgrow the free data allotment if you decide to back up all your files. If you are deep into security, you may prefer more complex (and expensive) services.

I've tried several of these services and settled on Dropbox for my professional work and Apple iCloud for photos. For less than $100/year, Dropbox provides two terabytes of data that accommodates everything I have ever written, every document required for hospital credentialing and licensing, my reference library of thousands of journal articles, my best underwater films, and a large number of miscellaneous files. Hard copies of all of these would fill a small room and be tough to lug around from assignment to assignment. Dropbox automatically synchronizes every file to my three desktops (one in Rhode Island, two in Tennessee), two laptops (one in my home office, one in my travel bag), iPad and iPhone.

Cloud Security

Like your laptop, access to cloud accounts requires a password. Two-step verification adds an additional layer of security and requires a smartphone (see below). Of course, no security system is perfect. Even reputable cloud storage companies, like Dropbox, have undergone data breaches.

One locum tenens physician recommends an additional safety step. He adds an encryption application, "Boxcryptor," to protect Dropbox files. In principle, even if his cloud files were hacked, they would be unreadable. A number of companies offer applications for this purpose. As security concerns continue to grow, it's probably only a matter of time before I adopt an encryption program as well.

True Story

Computer crashes, unfortunately, are not rare. The late Margot Kidder who played *Daily Planet* reporter Lois Lane opposite Christopher Reeve's

Superman, famously lost three years-worth of her autobiography due to a computer virus. That traumatic event contributed to a nervous breakdown (Poole 2002).

I've experienced two computer crashes, albeit with somewhat less drastic sequelae. The first was many years ago before cloud storage became commonplace. All the data in my otherwise trusty Dell computer disappeared overnight. Over several days, I frantically reconstituted most of it from a haphazard collection of hard drive backups.

My second computer crash occurred more recently in an aging Apple laptop. One day I pressed the power button and nothing happened. (Yes, it was plugged in.) Desperate calls to the Apple 800 number ensued. The inexplicably calm representative explained that the computer had outlived its useful life and was not worth salvaging. The data might be retrievable, but the process would be expensive and not guaranteed.

The solution? I purchased a new computer, logged into my cloud account, and took a leisurely nap while all my files reconstituted. A few hours later, I was back at work without losing a single bit of data.

A nuisance? Yes. A catastrophe? No. Moral: Sign up for a cloud account and backup your data.

Physical Backups

If you worry about cloud dependability (or security), you can utilize high capacity external hard drives as additional backups. Small, portable hard drives are useful to back up specific files, like slide presentations. These go by a variety of monikers such as data stick, flash, jump, key chain, memory unit, pen, and thumb drive. Rugged and waterproof hard drives suitable for road warriors are available.

You may wish to encrypt these hard drives in case you lose them. Otherwise, your data will be accessible to anyone who plugs the drive into their own computer.

Physical backups are great complements to cloud storage. Both can fail, so it's best to have at least one of each.

Sublime Synchrony

The value of synchronized cloud storage can't really be appreciated until you use it. For example, as a compulsive writer, I might work on my laptop during the flight home, edit on the iPad before bed, switch to the desktop the following morning, then edit some more during lunch on my personal computer at the hospital. In pre-Dropbox days, I painstakingly emailed documents to myself in order to have the latest version available for editing on whichever device I was using. A vexing lack of clarity regarding which version was actually the most recent frequently ensued.

Cloud storage eliminates that awkward extra work. When you save a Word document, Excel spreadsheet, or PowerPoint slide set, the file is backed up and updated on all devices instantaneously. As your devices multiply, this synchronization feature becomes even more treasured.

If you work without internet, say on the airplane or deep in the bowels of the hospital, the synchronization feature can't function. Once you're back in a Wi-Fi zone, your laptop will automatically connect and synchronize files to the cloud.

Smartphone

If you are a physician less than 35 years old, read no further. There is a 99.99% probability you have a smartphone and know how to use it. However, if you are not yet in possession of an iPhone or equivalent, it's time to get started.

In 2007, Apple introduced the iPhone. Refusing to be one of those pathetic people texting their way into fountains, I vehemently resisted this technology. Who needed one of these newfangled gadgets anyway?

In late 2014, I signed up for a "7 on, 7 off" neurohospitalist assignment that required weekly flights from Providence, RI, to Minneapolis, MN, via Chicago, IL. As O'Hare airport in Chicago is famous for delays, it was obvious this travel schedule would engender a lot of downtime. I'd heard of smartphone applications (apps) that might mitigate travel woes and reluctantly bought an iPhone 6.

It took about five minutes to realize that the smartphone purchase was the right decision. Not having a smartphone at the airport would be like traveling to the North Pole without gloves. You could do it, but why?

Smartphone and Business

Smartphones offer the ability to manage banking, credit card, mortgage and online health accounts. While sipping coffee at the airport, you can quickly access your bank and see whether your beloved staffing agency has deposited last week's paycheck.

Smartphones offer immediate access to personal and business email. Inexpensive apps allow you to scan and catalog business cards from fellow travelers and new colleagues or scan documents and file PDFs in the cloud. Visual voicemail saves the bother of listening to lengthy phone messages. Last, but not least, you have a telephone!

As a traveling physician, your landline has lost much of its relevance as a "home number." I stubbornly clung to mine for years, as it seemed a "real business" ought to have a permanent phone. Reluctantly, I cancelled it a few years ago. I haven't missed it for a second!

Smartphone and Medicine

A smartphone is an essential addition to your trusty black bag. It can provide an army of apps that assist with day-to-day medical practice.

For example, to write prescriptions I used to rely heavily on the Physicians' Desk Reference (PDR), a weighty compendium of drug information found in every clinic and hospital. Now there's a PDR app. Better than the original, it's always up-to-date and in your hand. Epocrates is another popular drug referencing app.

During stroke codes, one of my apps calculates tPA dosage. For clinic patients, an algorithm weighs the advantages of aspirin as an antiplatelet agent against the risk of bleeding. Other apps calculate creatinine clearance, anion gaps, phenytoin doses and stroke risk from atrial fibrillation. One free app even offers Ishihara plates to test color vision.

UpToDate is a great point-of-care resource for the latest treatment guidelines. I regularly read medical news and reviews on Medscape.com, which also offers a drug information and drug interaction app. When stuck in traffic, I listen to medical podcasts on ReachMD.com. (Disclosure-I write for Medscape and host interviews on ReachMD.)

New medical apps appear regularly. Reviews can be found at imedicalapps.com and other sites.

Smartphone Tips

Cell phone data plans constantly change. I pay for an "unlimited" plan, which eliminates worry about call minutes, megabytes of data or number of texts. As you are likely to become quite dependent upon your smartphone, I suggest a similar approach. You have more important things to do than track your cell phone usage. If it turns out you don't use your cell phone much, you can always downgrade your plan. Remember, the portion of your smartphone bill that applies to your locum tenens work is tax-deductible (Chapter 14).

Don't pinch pennies on your phone's memory either. Upgrade to the maximum you can afford. Insufficient memory hampers the function of your expensive new pocket computer. Would you put a four cylinder engine into a brand new Ferrari?

Your smartphone won't work with a dead battery. Buy an extra charger and stow it in your travel bag. Consider a powerpack to keep your phone alive during cross-country flights or long hospital days.

New Contacts

When you start an assignment, you will meet plenty of new people. Each job results in a long list of email, telephone and other contact information. You will interact with some of these acquaintances frequently, but others you may speak to rarely, if ever, again. When problems or questions come up on assignment, remembering whom to call at each facility can be challenging.

One organizing tip is to enter the hospital or clinic as a prefix to the person's name. For example, all administrators and physicians I met at the Mayo Clinic in Phoenix, AZ, are listed under "Mayo." Similarly, all contacts from Angels Neurological Clinics in Boston, MA, begin with "Angels." After the name, I add that person's role, such as HR, MD, or payroll. When returning to an institution for a second assignment, this strategy proves valuable. In addition, it gives others the impression you have a fabulous memory!

Smartphone Preservation

Your smartphone will become your new best friend while traveling. Do not skimp on a protective case because you *will* drop it. LifeProof and Otterbox are two well-known brands.

When I slipped and fell on an ice-covered sidewalk in downtown Minneapolis, my phone survived thanks to a heavy duty LifeProof case. The fall was sufficiently dramatic that bystanders rushed to assist. Although I was severely bruised (and embarrassed!), the phone survived unscathed.

Email

As a locum tenens physician, you will likely spend a lot of time away from home. Email will become an essential link to colleagues, family and friends. In addition, you will have work emails onsite. Here's advice about the latter.

Imagine you've just arrived at your new locum tenens assignment. During orientation, the institution assigns you a snappy new email address. It has an ".edu" or ".org" suffix, your name, and the institution's name or acronym. It looks very professional. You can't wait to use it.

Do *not* use that email unless you absolutely must. Why? Here are three reasons:

1. Your privacy is at risk (Chapter 10). Did you know your new email is not really yours? It is subject to "auditing" and "monitoring," a euphemistic way of saying management can read your email. In addition, at the drop of a hat, your email access can be withdrawn without your consent. Think about that.
2. Email is a time sink. You are getting paid to see patients, not respond to emails. When your shift ends, you don't want to be tapping away at the keyboard. Wouldn't you prefer to be sampling local restaurants, seeing the sights, or getting some rest? That's why you picked locums!
3. One strategy to keep your inbox from overflowing with emails is to avoid sending any. The more emails you write, the more responses you'll receive. Respond only when absolutely necessary. In addition, the more active your email, the more spam you'll receive. The

mathematical relationship of sent emails to spam is exponential. Had Sir Isaac Newton lived in our time, this would have been his 4th law of physics. You already have enough spam in your personal Yahoo, Gmail, or Hotmail account. You don't need any more.

If you restrict your institutional email to business, when you see an email in your hospital account, you know it's work-related and important. Instead of the 200 or more emails a day in my Yahoo account, my institutional account receives only a few. I respond to these promptly.

The other side of the coin concerns your personal email. Do not use it for patient care! Your institutional email maintains strict encryption protocols that comply with the U.S. Health Insurance Portability and Accountability Act (HIPAA). Your personal email probably doesn't. Yahoo and Gmail, for example, are not HIPAA compliant.

There's also the unfortunate reality that your emails may get hacked, often through no fault of your own. Email malware, phishing, trojans, and viruses have become more common and difficult to detect. If your personal email gets hacked and patient information is exposed, you increase the likelihood that the U.S. Government Office of Civil Rights (OCR) will come knocking at your door.

HIPAA violations can result in civil fines and criminal penalties. You don't want either. Separate your work and personal emails, both for your own privacy and to preserve patient confidentiality.

Fax

Desktop fax (facsimile) machines have been around since 1948. Digital faxes first appeared in the 1960s. These days, fax has largely been replaced by transmitting scanned documents through email. Although fax use is fading, many clinics and hospitals stubbornly cling to this ancient technology.

A personal fax number does hold a certain appeal. It looks impressive on office stationery and lends the impression of "a real business." It may also facilitate communication with other professionals, such as your attorney and accountant. I receive about one fax a week to my digital fax number (see below) from my accountant, whose office staff still depend on their antiquated fax machine.

For the itinerant locum tenens physician, a traditional fax machine offers limited usefulness. Back in the day, I had a thermal printing fax machine and paid for a dedicated fax line. When I was home, it was useful. But after an assignment, I would often return to reams of fax paper on the floor, or worse, a jammed machine and no faxes at all. I trashed that big, black machine years ago.

However, I did replace it with a digital fax number that allows me to send and receive documents from my personal computer. Faxes materialize as PDF files. That digital fax has proved valuable in many travels over the years.

Nowadays, documents can be readily scanned and emailed, a process generally faster than faxing. Desktop scanners and even smartphones can produce respectable scans. Electronic signatures on PDF documents are suitable for many contracts, eliminating the need for signed, faxed copies.

If you run your own private practice, a HIPAA compliant fax could prove useful. But as a locum tenens physician, a private HIPAA compliant fax is superfluous. Why? Because you will keep all protected health information (PHI) on the hospital system and not your personal devices.

Doximity.com provides a free digital fax number as an incentive to join its online physician community. This works on a computer, iPad, smartphone and other devices. (I mention Doximity.com because it's the only source I've found for a free HIPAA compliant fax number for physicians.)

Multiple smartphone apps emulate physical fax machines. If you don't like free services, you can get a paid fax number from eFax or other companies. Limited free fax service is available from the Faxburner app.

Many hospital and other credentialing applications request your fax number. You will suffer no adverse consequences if you don't have one.

Not for Patient Data

As discussed above, to comply with HIPAA regulations, you will never transmit patient data in your personal email or fax. Similarly, none of your gadgets such as laptop, smartphone or tablet should ever store patient data.

The same advice applies to your cloud service. Several cloud storage systems claim HIPAA compliance (e.g., Box, Dropbox Business Edition, OneDrive for business). But even if your cloud storage is HIPAA compliant, why store patient data? There is never a good reason to use personal accounts for patient information, so it's a moot point whether your cloud service, email or fax are HIPAA compliant or not. If you must have access to patient data off-site, ask your hospital or clinic to provide it through a secure electronic medical record (EMR) portal (Chapter 16).

The federal government treats HIPAA violations very seriously. For example, in 2017, the OCR fined the Children's Medical Center in Dallas, TX, 3.2 million dollars for the loss of an unencrypted Blackberry, iPod, and laptop that contained patient records. The government levied the hefty fine even though no harm came to any patients as a consequence of this security breach.

Your hospital spends millions of dollars a year on its corporate compliance team to secure emails and maintain confidential patient data in its EMR system. That's their job, not yours.

Stalked by a Pharmacy

Using your devices for clinical work creates another problem. You may find that patient information, such as a prescription refill or lab result, turns up after you've left your assignment. This happened to me when a pharmacy somehow got hold of my personal fax number and repeatedly faxed prescription refill requests. (I think they found the number in a physician fax directory.)

Refill requests are always annoying, but this situation created an awkward, time-consuming dilemma. I was already out of state and at another assignment.

Failure to renew these prescriptions could be viewed as patient abandonment, with severe consequences. On the other hand, since I no longer had staff privileges, continued patient care would be inappropriate if not illegal. Each time it happened, I had to call the pharmacy, explain the situation, and redirect the refill request. The hours "on hold" and speaking with various pharmacists were frustrating and uncompensated.

One of the beauties of locum tenens work is when you're done, you're done. Do everything you can to keep it that way.

Fast Wi-Fi

Your mobile office requires internet access. Consider a fast connection (>15 Mbps) for your home office. The extra cost will more than pay for itself with added efficiency. Further, if you subscribe to Amazon Prime, Hulu, Netflix or other media services, your programs won't be subject to "buffering" and you'll be able to enjoy them.

Wi-Fi at airports, coffee shops, hotels and other public places may expose your personal information to hacking. If you are concerned about the security of your data (and you should be), consider a Virtual Private Network (VPN). A VPN provides a kind of virtual "tunnel" that protects your data within a public Wi-Fi network. Since you will access countless different Wi-Fi hotspots in your travels, a VPN subscription might prove prudent for your laptop and other devices. Popular VPNs include ExpressVPN, NordVPN, CyberGhost and others.

Pick One

All essential home office tools have a steep learning curve. For example, there are many cloud storage options, such as Apple iCloud Drive, Box, Dropbox, and Microsoft OneDrive (see above). For a while, I juggled files between them, trying to take advantage of their free storage allotments. This turned out to be a poor use of my time. Ultimately, I subscribed to Dropbox. Similarly, there are several personal email options (i.e., Google, Hotmail, Yahoo). I had several, but now rely on Yahoo nearly exclusively.

For word processing, spreadsheet and slides, you can choose from several program sets; Apple (Pages, Numbers, Keynote), Microsoft Office (Word, Excel, PowerPoint), Google Docs (Docs, Sheets, Slides) and others. Microsoft Office is the most widely used and will cause the least number of compatibility problems when you log in from different computers. Choose one set and master it.

Hotel Business Center

If you forget or lose your laptop, many hotels host a business center with a desktop computer and printer. Unfortunately, these are not always well-maintained. Paperless printers and frozen computer screens are common. In addition, fees may be required. If they work, these tools can be very convenient for simple tasks like checking email or printing a boarding pass. When you leave, take care not to leave passwords logged into your email or websites.

Old Fashioned Way

Your new hospital or clinic may not provide anything more than a shared cubicle for your office. Work space for managing personal paperwork and correspondence may default to a flimsy hotel desk or night table.

In you insist on proper office equipment like a desktop computer, land-line and fax machine while traveling, check out Regus.com. Regus offers full service office rental space and meeting rooms in many locations all over the world. Rentals can be as short as a day. I've never used them but have been tempted when I needed a professional workspace other than my hotel desk, particularly when boisterous guests occupied the adjoining room.

Recommendations

Buy the best laptop and smartphone you can afford. Sign up for cloud stor-age. Restrict that sexy work email for official business. Forget the fax. Settle on one office productivity software suite. Never use your cloud, email, fax, smartphone or tablet for HIPAA protected patient information.

CHAPTER 16

Electronic Medical Record

Doctor vs. the EMR

In the latter part of the 20th century when I was in private practice in North Carolina, several senior physicians refused to log on to the hospital computer. They complained and stalled, desperately hoping they could continue their old ways until retirement.

In the 21st century, computer literacy is no longer optional for physicians (and anyone who doesn't live in a cave). Private practices that continue to operate with paper charts are a shrinking minority and financially penalized by the Centers for Medicare and Medicaid Services (CMS).

All major hospitals use some kind of electronic medical record (EMR). As a locum tenens physician, you must become familiar with one or more of these systems. All are complicated, difficult to learn and awkward to use. To add to the misery, computer screens physically obstruct patient-physician interaction.

Struggling to care for patients while battling an EMR may contribute to physician burnout (Sinsky et al. 2016). My own experience with Epic may prove enlightening, if not entertaining (Wilner 2013).

EMRs are not without benefits, but these come with downsides. For example, I do enjoy computer access to neuroimaging, but miss morning radiology rounds (Wilner 2010). For better or worse, the EMR has become an integral part of patient care.

EMR and EHR

EMRs are big business. *Medical Economics* listed 100 Electronic Health Record (EHR) companies desperate to host your patients' medical information (Medical Economics 2013). Epic is the most commonly used EMR, accounting for more than a third of the market, with nearly 200 million patient records. Additional EMRs you may encounter include Allscripts, Cerner, eClinicalWorks, NextGen and others.

Technically, EMRs are digital versions of a patient's chart, while EHRs are broader, connecting digital charts to different health care systems. You will hear both acronyms used synonymously. (I wager a free copy of this book to anyone who can find two practicing physicians able to articulate the difference between an EMR and EHR.)

Resistance is Futile

Many physicians have opined that the EMR's prime function is to enhance billing and decrease productivity. While you may ultimately find the EMR helpful in patient care, it is a rare physician who enjoys using one. Clicking boxes all day does little for the soul.

Nonetheless, the more skillfully you operate this modern tool, the more attention you can devote to your patients. EMR proficiency will amaze your colleagues and get you out of the hospital or clinic on time.

Bait and Switch

Every hospital I worked at during the last ten years already had an EMR, was transitioning to an EMR from paper charts, or was in the process of switching to a different EMR.

It is not unusual for hospitals to change EMRs. One Connecticut hospital struggled for years to implement a McKesson system without success, squandering thousands of man-hours and millions of dollars in the process. When the hospital was finally swallowed up by a larger health care system that used Epic, the administration promptly abandoned the bungled McKesson implementation. I rue the time spent in mandatory McKesson seminars and wasted hospital dollars that could have been better spent to relieve congestion in the overcrowded ER and improve patient care.

Recently, I worked at a hospital in Phoenix, AZ, which used Cerner. Having already learned Epic, the transition wasn't too difficult. Mind you, IT consultants spent many hours to walk me through the countless clicks required to reconcile medications and refill prescriptions. Towards the end of my locums assignment, I learned that this large hospital plans to convert their medical record system to Epic next year. If only they had switched sooner, I could have skipped the Cerner experience altogether.

As a further irony, my new hospital uses no EMR system at all for inpatient progress notes. Doctors scribble in the chart like the good old days. The administration claims it is implementing Cerner, but missed its own deadline months ago...

Patient Portal

An increasingly popular EMR feature, spurred by the 2014 CMS Electronic Health Record Incentive Program, is the online patient portal. Portals allow patients to review their medical records, check lab results, request prescription refills, and send secure messages to physicians and other health care providers. The CMS Incentive program requires health care providers to employ patient portals in order to qualify for Stage 2 "meaningful use." Failure to achieve meaningful use results in a reimbursement cut. Consequently, medical practices have scrambled to set up patient portals whether they want them or not.

On the upside, patient portals provide 24/7 convenience for patients. In my less preferred role as patient rather than physician, I appreciate the opportunity to look up my lab results without having to engage in telephone tag with the doctor's office.

For physicians, portals create yet another flow of information from patients that requires a response, a definite downside. I dread checking my EMR inbox as it often represents an hour or more of relatively unsatisfying, additional work.

Typing vs. Dictation

Your EMR may offer the option of dictating into specific text fields rather than typing. If so, workstations may be equipped with a special handheld microphone. The most common dictation system for medical charting is "Dragon" by Nuance. Ask for training if it's an option you want to pursue. At one hospital, I received three hours of CME credit for Dragon training.

Many physicians have become dictation virtuosos through years of practice. When I was a medical intern at the Long Beach Veterans Administration Hospital in Long Beach, CA, it was the intern's job to dictate every discharge summary. Given that some patients had thick paper charts from months of hospitalization, this was often a colossal task. When I left the VA to become a medical resident at Los Angeles County Hospital, Los Angeles, CA, it was the *resident's* job to dictate discharge summaries. Those three years of extensive, but not entirely welcome, dictation experience created a solid foundation for future dictations.

Dictating to an EMR, however, is different than dictating to a living transcriptionist. You must have a pretty good idea what you are going to say before the words leave your mouth. Such forethought does not always come naturally. If I don't compose my thoughts before speaking, the computer can't handle the jumbled output. Time I might have saved dictating rather than typing is lost in a tangle of edits. Thinking before speaking is a rare but acquired skill that improves dictation and may yield additional benefits.

Typing is another important secretarial skill for locum tenens physicians. Thankfully, I learned to touch-type in high school. My medical journalism career subsequently provided plenty of practice. Knowing how to type comes in handy as today's physicians spend twice as much time in administrative tasks than seeing patients (Sinsky et al. 2016). If you are a two-finger typist, consider a typing class. Free tutorials are available on the internet.

Expertise in both dictation and typing positions you for success in the modern physician's role as medical data entry technician. Depending upon the situation, you may choose to dictate or type. Usually, I type most clinical notes. However, if I have a long story to tell, such as the medical history of my Mayo Clinic patient who fervently described her 20 years of "four types of dizziness," dictation mode is a blessing.

Mastering an EMR is no small task. Large hospital systems offer classes and EMR specialists. Take advantage of them. I have sat through hours of classes on Epic and Cerner. Despite this training and assiduous attention to detail, my proficiency remains suboptimal. Like learning a foreign language, there is always more to know. Even experts who teach these systems possess significant knowledge gaps. But there is a magical point where you know enough to get by, and you must reach it as soon as possible.

Remote Access

The capacity to remotely access your hospital or clinic EMR will improve your productivity. This will allow you to review a stat CT scan, for example, or monitor a critically ill patient's electrolytes from afar. If you think this capability will be beneficial, request remote access. Once you obtain permission, bring your laptop to the IT department. There's nothing like a doctor standing in front of an IT technician to get priority in the queue.

The downside of remote access is that it makes it more difficult to emotionally disconnect from work. But if remote access spares one middle of the night trip to the hospital, it is definitely worth the effort to get the necessary clearance and complete the software installation. I often log on and check the patient census over breakfast to eliminate any big surprises when I get to the hospital.

Recommendations

Learning one EMR is bad enough, but locum tenens physicians may have to master several. Find out which EMR you will use at your next assignment. Insist on EMR training during orientation. Learn to dictate and type. Request remote access.

CHAPTER 17

Continuing Medical Education (CME)

> *The art is long, life is short, opportunity fleeting, experiment dangerous, judgement difficult.*
>
> Hippocrates

Lifelong Learning

All physicians recognize the importance of lifelong learning. Successful medical practice requires mastery of an ever-expanding database of scientific information, as well as finely-honed interpersonal skills and medical judgment.

Many physicians revel in opportunities to learn about new medications, procedures, even new diseases. Unfortunately, state mandates for continuing medical education (CME) impose a bureaucratic burden on this necessary and otherwise pleasurable endeavor.

State to State Differences

While each state sets its particular CME requirements, 40-50 hours every two years is the norm. Massachusetts was an outlier and used to require 100 hours every two years, but decreased its requirement to 50 hours as of January 1, 2018. South Dakota is another outlier. It leaves physicians to their own educational strategies and doesn't require CME.

State medical boards post CME requirements on their websites. An overview of each state's CME appears at www.fsmb.org.

The Fine Print

To further complicate the lives of physicians who hold more than one medical license, states may require CME that covers specific topics. For example, of its 40 hours, Rhode Island requires at least four hours related to end of life/palliative care, ethics, opioid pain/chronic pain management, and risk management. Not to be outdone, Massachusetts requires two hours of study of its own Board of Registration rules and regulations, as well as ten hours of risk management. Of its 50 hours of CME required every two years, Connecticut demands one hour of CME in each of the following; behavioral health, cultural competence, domestic violence, infectious disease, risk management and sexual assault.

There is more inconsistency than logic in state-specific CME requirements. For example, Nevada, among the top five states for suicide, rightly requires its physicians to obtain two hours of CME in suicide prevention and awareness. But Washington State, which is not even among the top ten states for suicide, requires six state-sanctioned suicide prevention CME hours.

As you expand your quiver of state licenses, the work of maintaining required CME increases. At one point I had ten active state licenses, which forced me into "spreadsheet mode." You can guess where I stand on the proposal for a national medical license with a single CME requirement.

Individual state medical boards have no incentive to standardize their respective CME demands. Unfortunately, the much-heralded Interstate License Medical Compact (ILMC) does nothing to ameliorate the challenge of complying with multiple state CME requirements (Chapter 9).

Audits

As a self-employed physician, tracking CME credits is your responsibility. License renewals typically require a mere attestation that CME credits are up to date. One can be lulled into a false sense of security by this relatively lax standard. Unfortunately, I had a wake-up call when my CME records became the subject of a random audit by my home state of Rhode Island. They insisted on every CME certificate from the prior two years. That episode occurred in the days when I casually stuffed documents into a metal filing cabinet. Lots of fun complying with Rhode Island's request!

Record Keeping

There is no single agency that tracks CME credits, leaving that secretarial function to you. Since that humbling CME audit, I keep a spreadsheet of CME credits and update it regularly.

The spreadsheet includes the following necessary information: Date, Title, Sponsor, Accreditation Agency, Location, and Hours. If a CME course satisfies one of the special state categories, like pain management (a very "in" topic), I put an asterisk by the title on the spreadsheet. In addition, I scan every CME certificate (or download the PDF) and file it in a Dropbox folder (Chapter 15). With this system, when a state or hospital credentialing committee requests my CME record, I effortlessly comply with a few mouse clicks.

CME Categories 1 and 2

All the above state requirements refer to Category 1 CME credits. According to the American Medical Association (AMA):

> Category 1 credit represents that the physician has participated in an educational activity, and completed all requirements for such an activity, that is expected to "serve to maintain, develop, or increase the knowledge, skills, and professional performance and relationships that a physician uses to provide services for patients, the public or the profession."
>
> Category 1 credit is the most commonly accepted form of CME credit for physicians and is also the basis for receiving the AMA Physician's Recognition Award (www.ama-assn.org).

States may also recognize Category 2 CME. Category 2 credits are self-designated. I have never bothered with Category 2 CME credits as they do not satisfy state CME requirements for license renewal.

Cost

A burgeoning industry has sprung up to provide physicians with CME. The Accreditation Council for Continuing Medical Education (ACCME) lists over 1,750 organizations that offered more than one million hours of accredited CME in 2016 (www.accme.org).

No centralized listing of certified CME programs exists. However, if your state requires CME in a particular category, an online search should reveal a source. State medical boards may offer live seminars or on-demand web courses to fulfill their own requirements. For example, Washington State's requirement for six CME hours of suicide prevention is free on-line. Other state sponsored courses may require a fee (conflict of interest, anyone?).

Annual medical conferences are an easy way to obtain in-person CME. If you work in a hospital, Grand Rounds traditionally offers a free lunch and an hour's credit. Unfortunately, traveling to conferences consumes time and money, and lunchtime lectures are falling by the wayside.

The internet has spawned many companies that offer free online CME. Three such resources I utilize regularly are Medscape.com, Medpagetoday.com, and ReachMD.com.* These CME providers offer a variety of programs and tally your CME. If you are a visual learner, TheDoctorsChannel.com specializes in free video CME.

Certain medical journals, such as the *Annals of Internal Medicine* and *MayoClinicProceedings*, offer CME. Academic-minded physicians may earn CME credit when publishing journal articles or serving as medical reviewers. The AMA lists multiple other avenues to obtain CME (www.ama-assn.org).

Most salaried physicians receive at least $1,000 and one week off to obtain CME. However, as a locum tenens physician, you must fund your own education. While staffing agencies do not provide a CME allowance, at least two offer free CME courses. I have taken many of these, and they are high quality. Your medical malpractice carrier may also offer free CME on risk management.

Sometimes, a required CME course is difficult to locate. In these instances, I've resorted to purchasing courses from VLH.com, which offers all state-required CME. Other online companies fill this niche as well. If you must pay for CME out of your pocket, these costs are tax-deductible.

Recommendations

Track CME on a spreadsheet and scan certifications into an online folder. Satisfy CME requirements for each state license.

*Disclosure: I've written and hosted CME programs for Medscape.com and ReachMD.com.

CHAPTER 18

Travel Tips

Travel

Travel isn't always required for a locum tenens assignment. You may find the perfect one in your own backyard. But for many physicians, travel becomes part and parcel of *The Locum Life*. According to a recent survey, 89% of locum tenens doctors prioritize their assignment by location, 67% by pay rate, and 60% by length of assignment (Staffcare 2017). Travel will likely be necessary in order to engage a desirable assignment in your specialty.

The frequency and extent of travel can vary tremendously. For example, one of my assignments required an hour and a half drive to the office two days a week. In another, American Airlines shuttled me from Arizona to Minnesota once a week in economy class. A longer plane ride carried me between South Dakota and Rhode Island. For my stint at the Mayo Clinic Phoenix, I gave up the frequent travel and relocated to sunny Arizona for a year. It's all up to you.

Efficient Travel

As a locum tenens physician, travel to and from the workplace may swallow a significant portion of your free time. Staffing agencies will reimburse major travel expenses, but will not compensate you for travel time. All those hours in the car and airport are on your own dime. In my biased view, this is a near fatal-flaw in the locum tenens reimbursement model. Two of my assignments required weekly travel for their 7 days on/7 days off schedule. That's two days of travel for every week of work!

In order to compensate for lost income en route, I rely on my laptop to keep productive (Chapter 15). With a reliable computer and cloud access, the world is your office. Email can be answered, credentialing applications filled out, journal articles reviewed, and receipts tallied. I've even written blogs and prepared PowerPoint presentations on the airplane tray table.

All major airports offer Wi-Fi, and an increasing number of aircraft possess Wi-Fi capability. Airport Wi-Fi is often free, but once airborne, Wi-Fi requires a fee. One little-known Wi-Fi hack is free text messaging and an hour of internet surfing per flight for those with a T-Mobile phone plan. I use this one a lot.

If you would rather read a novel or watch a movie, or you are one of those lucky people who can snooze on the plane, it's your time to spend as you wish.

Comfort

Modern travel has become uncomfortable. Long lines at check-in or security, cramped seating in economy class, excessive air conditioning and fast food all contribute to travel-related fatigue and frustration. Since it's important to be at your best once you arrive at an assignment, any measures you can implement to lessen the discomfort are worth considering.

First class travel offers larger seats, better meals, and a far more comfortable experience. Staffing agencies will reimburse only for economy class, but you can upgrade at your own expense.

I've seen passengers travel with eyeshades, their favorite pillows, blankets and slippers, anything to cushion a long plane ride. One locum tenens physician brings a generous supply of sanitizer to decrease the germ count on his tray table. Noise-cancelling headsets diminish the constant rumbling of jet engines, crafting a more tranquil environment for reading or sleep. (These expensive headsets don't offer much insulation from a crying baby in the next seat, however.) An iPad or other tablet can provide welcome distractions in the form of books, films, music or solitaire. Some of these travel accessories may satisfy the "supplies and materials" internal revenue service (IRS) category of deductible expenses. Consult your certified public accountant (CPA) (Chapter 14).

Trip Planning

Major staffing agencies provide in-house travel agents to book flights, hotels and rental cars. In order to travel the most direct routes, I research the best flights and forward them to the agency. It's less work for the agent and better flights for me.

Ask which hotels or other living arrangements are available. You may have options. As for rental cars, the agency likely holds a contract with one major company. If you'll be doing a lot of driving, ask for an upgrade to a larger car. It may cost only a few dollars more per day, and the agency may pay for it.

You are eligible for "points" for all these travel necessities (see below). Before long, you will become a "preferred" customer, which entitles you to perks and a higher level of service.

Gadget Planning

Your phone, tablet and laptop comprise your mobile office (Chapter 15). Charge them before you leave home. Although some modern commercial aircraft have electric outlets next to each seat, many don't. I traveled on a first-class upgrade a few weeks ago in an older jet, and there was no outlet to revive my phone!

Before you leave on a trip, check whether any software updates are pending for your devices. Updates are best done at home while your device is plugged in and you have fast internet. Public Wi-Fi may not allow software downloads.

Before I travel, I download a few movies to my phone or tablet. These come in handy when the internet is sketchy and movies download slowly or not at all.

Packing

I always travel with the same suitcase and carry-on. I hate doing laundry, so I pack a full week's worth of pressed hospital clothes, a couple of jeans and polo shirts, and a few shorts and t-shirts for workouts. My carry-on has a duplicate set of chargers for an ever-increasing number of gadgets (see above). This way, I don't have to round up chargers from home, and I'm sure not to leave home without them. Women may also wish to keep a duplicate set of essential cosmetics in their carry-on, as well as other items they can't possibly live without.

Delays and Cancellations

Once you've checked your luggage, obtained a boarding pass, and found the gate, you should soon be on your way. However, flight delays, missed connections, and cancelled flights have all hindered trips to work. According to the Bureau of Transportation Statistics, flight delays on my favorite airline occurred more than 18% of the time and averaged more than an hour. A little over 1% of flights were cancelled (https://transtats.bts.gov).

Zig Ziglar, the sales wizard and motivational speaker, famously took a cheerful approach to flight cancellations due to weather and mechanical problems. Rather than express annoyance, he celebrated the pilot's wisdom in cancelling the flight!

Not all of us have Mr. Ziglar's unflinching tolerance for disrupted travel plans, particularly if it means a late arrival for an assignment. One practical piece of advice is to keep personal essentials such as toothbrush, medications, and a change of clothes in your carry-on. If you fly frequently, you will need them.

Miles and Points

Airline frequent flier miles, hotel, rental car and credit card loyalty points obtained on the job belong to you. While an occasional flight or hotel stay won't amount to much, frequent plane trips and repeated hotel nights add up. If you can stick with the same airline, hotel chain, or rental car company, you'll maximize these perks. My American Airlines account has accumulated nearly 400,000 miles. That should get my family back and forth to the Philippines a few times, maybe even in business class.

Many internet resources rank frequent flier programs and credit cards with respect to points or "cash back." Now that I've accumulated sufficient frequent flier miles to last a while, I use a credit card that automatically returns 2% into my checking account. There may be better ones, but this one doesn't require any fussing or annual fee.

Smartphone and Travel

A smartphone at the airport is essential. This handheld device allows you to determine whether your flight is on time, check-in, choose a seat, get your boarding pass, find the gate, check the weather at your destination, and confirm your hotel and rental car. You will be among the first to learn of cancellations, flight delays and gate changes when text notifications appear on your phone.

For example, on the way to my first day of work in Sioux Falls, SD, the connecting flight never arrived. Stuck in the glacial customer service line with more than 100 other forlorn souls, I used my phone to search for another flight. While still stuck in line, I spoke to an airline agent and rebooked my ticket!

The clock on your smartphone automatically adjusts for time zones, so no more excuses for showing up late. In addition, every smartphone contains an alarm clock, so one less item to pack.

Smartphone cameras now rival or exceed the quality of many compact cameras. Unless you are a serious photographer who can't survive without a DSLR (you know who you are), you no longer have to pack a separate camera. Since I switched to a smartphone, my compact digital camera collects dust at home.

Smartphone and Family

On the road or at your destination, phone applications (apps) like Facetime, Skype and others allow face-to-face conversations. If you are on Wi-Fi, nearly ubiquitous these days, the cost is nil.

Text messaging allows succinct communication (i.e., "I've arrived. Call you later."). Instead of pecking away at the phone keyboard, you can dictate. Voice recognition is not perfect, but usually sufficient to get your message across.

Smartphone and Entertainment

You'll need some amusement on the road. Smartphones offer endless entertainment options including books, games, magazines, music and video. I read Kindle books on my phone during quiet lunch breaks. A large collection of free digital books and magazines are accessible courtesy of my local library. TV shows and movies from Amazon Prime, Hulu and Netflix stave off boredom on the treadmill. At night, white noise apps drown out street noise and smooth the transition to sleep.

Gadget Tracking

Do not lose your phone, tablet or laptop! During your travels, you must keep an eye on these expensive pieces of equipment that contain your business information and life's secrets. Print your name and address in big letters on your gadgets in case some civic minded person finds the lost item. Several companies make luggage tracking devices that might also make a good investment for your electronics.

Home Office

With your laptop, smartphone, and/or tablet as well as documents stored in the cloud, you are prepared to handle all demands of business, communication, education, even entertainment while away from home. Your office travels with you.

But one anachronism of the modern world stubbornly refuses to go away, "snail mail." Bills, checks, license renewals, important missives and packages still arrive the old-fashioned way. Unopened, they sit in your mailbox or languish on the front porch until you return.

There are solutions. First, sign up for "autopay" for recurring utility bills. You don't want to return home and find the electricity or gas cut off because months of bills went unpaid.

Second, select "paperless" for all bank and brokerage statements. You can receive email notifications and read them at your leisure.

Third, forward your U.S. postal mail if you are going to be away for any length of time. It can be done easily online at USPS.com in less than five minutes and costs one dollar.

Fourth, also at USPS.com, sign up for "Informed Delivery." This free service sends you an email with photos of snail mail due to arrive in your mailbox. It's a great "heads up" something important is on its way.

Finally, ask a trusted neighbor to look after your mail and home. You will not be able to foresee every eventuality. These "boots on the ground" will be able to gather up packages and address emergencies such as break-ins or storms.

Home Security

A single security camera with smartphone access surveys my home when I'm away. It streams live audio, video, and stores a week's worth of recordings for free. It sends an alert if anyone, like the cleaning service, enters my place. The camera I use is made by Netgear, but there are many brands to choose from. During long assignments, it's comforting to turn it on and remind myself that I have a home…

Take Care of Yourself

Many "road warriors" find living away from home detracts from efforts to eat nutritious meals, exercise, and get adequate rest. According to nutritionist Kathy King, RDN, LD, "You have to want to be healthy. That means you set boundaries on your behavior and you plan ahead when you travel and live away from home. I seldom find lack of lifestyle knowledge is the reason why physicians don't make good choices" (personal communication).

Maintaining a healthy diet and exercise routine can be a challenge while living in a hotel room. It helps if you insist that your health is not a luxury but rather a requirement. Remember, if you don't take care of yourself, no one else will.

Most hotels offer basic gym facilities, if not elaborate ones. The trick is showing up. Figure out the best time to exercise. (I used to exercise after work until I realized it interfered with my sleep. Now I lope along on the treadmill or elliptical for 30 minutes or so before breakfast.)

Long walks or jogging around the neighborhood offer interesting insights you would never get by car. Before you head off alone, check with the locals to see whether it's safe.

If your assignment lasts more than six months, you may be eligible for a rental apartment rather than a hotel room. While it will be more complicated to maintain an apartment, kitchen facilities that offer the option to cook your own meals may make it worthwhile. Discuss this possibility with your staffing agent.

In addition to physical health, one shouldn't neglect mental health while away. Stress relief strategies are important. One locum tenens physician advocated bringing a favorite pet to prevent both of them from becoming lonely.

Before you leave home, check whether you have any long-lost relatives or friends who live in the area of your new gig. Connecting with one person can open up a whole new world of acquaintances and activities you would otherwise miss.

Recommendations

Plan your travel. Try and enjoy it. Be productive on the road. Do not lose your gadgets! Eat right, exercise, get enough sleep, and take care of yourself.

CHAPTER 19

Strategies for Success

One Giant Step

If you've read this far, you're well on your way to a locum tenens assignment. You've learned what locum tenens is, it's pros and cons, whether it's right for your stage of career, the role of a staffing agency and whether or not you need one, how to garner hospital privileges and state licenses, protect yourself with professional liability insurance, negotiate compensation, operate as an independent contractor, organize your mobile office, master one or more electronic medical records (EMRs), obtain and track continuing medical education (CME), and travel comfortably, economically, and efficiently. Here are some additional tips, most of which I've learned the hard way, to insure success when starting a new assignment.

Qualities for Success

You have completed medical school, residency, and maybe a fellowship, unassailable testimony to more than a smidgeon of industry and intelligence. Your clinical diagnostic and therapeutic skills are satisfactory if not superb, and you constantly work to improve them. You reassure patients with a warm and gentle bedside manner. Your role model is William Osler, not Gregory House. With all that on board, which additional qualities and skills do you need?

It does take a special kind of person to succeed at locum tenens. Flexibility is key. More than one locum tenens physician has told me that they view themselves as a "guest in someone else's house." They are there by invitation. They often remind themselves that their stay is temporary. Their goal is not to redecorate or remodel, even when faced with a glaring need. Obstacles must be overcome and niggling workplace aggravations ignored. In this view, the job is to take care of patients and keep the wheels turning.

I agree. The "guest" analogy emphasizes the importance of social skills like listening and politeness. Coupled with clinical acumen and an old-fashioned work ethic, your success is virtually guaranteed.

Getting the Job You Want

Locum tenens recruiters search for physicians with specific qualities. An article published by CompHealth, a major locum tenens staffing company, itemized "7 skills hospitals are looking for in a physician" (Saylor 2016).

- Collaboration
- Communication
- Flexibility
- Life balance
- Putting people first
- Strong listening
- Time management

This list includes obvious necessary qualities for a locum tenens physician such as communication skills, flexibility, and putting people first. Time management skills are also essential when tasked with multiple patient care responsibilities in an unfamiliar environment.

The author also recognizes the importance of "life balance." He writes, "A good life balance creates a better employee, makes you more productive at work and helps prevent burnout. This is good for the physician and the facility" (Saylor 2016).

An article published by Barton Associates, another well-known locum tenens firm, "5 Qualities a recruiter looks for in a locum tenens candidate," listed similar qualities to those proposed by CompHealth (Cavanaugh 2017). The author wrote:

The perfect candidate is someone who books their schedule out two to three months in advance and has multiple licenses, and understands that a lot of times when they're going into locums; it's because there is an event that's happened. Sometimes they might be on multiple EMR systems, or the clinic might be a little short-staffed, or support staff might be a little low. They've got to be able to go in and handle that.

That sounds right. A locum tenens job exists because of an unexpected and difficult to satisfy clinical need. If it was the perfect position, there would be a long line of eager candidates to fill it. Such a desirable job would never come to the attention of a locums agency.

You must expect to walk into a clinical situation that is more or less a mess. Don't be surprised if your staffing agent downplayed the chaos or the workload. Like real estate agents, staffing agents tend to emphasize positives and downplay negatives (Chapter 8). This is where flexibility and strong interpersonal skills come in handy. Be prepared for the worst-case scenario. With luck, you'll be relieved it's not so bad!

I hasten to add, however, that at least three of my assignments have been wonderful. For all of them, the facility had a pressing need, and I was available. My workload was reasonable, if not light. Both patients and staff appreciated my presence and contributions. It was "win-win" for all of us. I would still be working those jobs, but the respective institutions recruited permanent physicians. My services were no long required. In one case, budget cuts eliminated my position. (Good thing I wasn't on the permanent staff!)

The 3 A's

Three fundamental qualities that will help you get hired and succeed once on the job are codified as the "3 A's of Medicine" (Murray 2017). In order of importance, these are availability, affability, and ability. In brief, if you show up, be nice, and get the job done, you'll succeed.

Availability is a necessary first step. Desirable locum tenens positions fill rapidly. When an agent calls with a new opportunity, the sooner you can commit, the more likely you will get the job. Multiple state licenses improve your eligibility.

The importance of affability, the second "A," cannot be overstated. As a locum tenens physician, you will always be the "new guy" (or gal). Patients and peers will quickly notice whether you are nice or not.

Frankly, I used to give affability short-thrift. For me, competence seemed a more crucial quality. However, I've since learned that affability is truly an important trait for success in every type of clinical practice.

Affable physicians listen well, take an honest interest in their patients (and peers), and are a pleasure to work with. It used to be that physicians with major character flaws could succeed thanks to outstanding clinical skills. Not anymore. The days of mercurial surgeons throwing tantrums and instruments in the OR are gone.

Luckily, affability doesn't take years of study, hard work, or sacrifice to achieve. Even physicians born with a defective affability gene can acquire the basics. Benefits outweigh the efforts.

Our modern, protocol oriented, politically correct, corporate medical environment has systematically marginalized prima donnas. In today's workplace, nice, marginally competent physicians fit in better than abrasive, exceptionally proficient ones.

On the job, affability comes in handy. Colleagues and nurses will enjoy working with you, and your patients will likely receive better care.

Affability helps cultivate collegial relationships. These will constitute a valuable resource when you need recommendations for future assignments.

Ability is the last of the 3 A's for a couple of reasons. First, no one cares how capable you are if you can't offer the first two "A's". Second, employers know you went to medical school, completed a residency, and obtained a state medical license. They assume you are competent.

Personally, I think ability should receive more attention. While all men are created equal, all physicians were not. Further, each physician doesn't maintain the same standards of attention, diligence and study required to consistently achieve clinical excellence.

Unfortunately, employers possess limited tools to assess your ability. There's an old joke, "What do you call a medical student who graduated last in his/her class?" Answer: A doctor.

Employers know whether or not you passed the boards, but not your score. Peer references are subjective and may be difficult to interpret, as they require reading between the lines. Surgeons may keep a registry of clinical outcomes, but no such data exist for nonsurgeons. In any case, clinical outcomes are often poor measures of competence, as they tend to reflect disease severity more than clinical expertise.

Your clinical ability will only be recognized in two extreme circumstances. One, you perform exceptionally well and reap admiration from peers. Two, you make frequent errors and leave wreckage in your wake.

Orientation

Insist on a thorough orientation on your first day. A brief tour should point out where you will work, the call room, emergency room, intensive care unit and doctors' lounge.

Before you treat patients, you will need nuts and bolts like an identification badge, parking pass, and computer passwords. Ask for training on the EMR and dictation system if you need it.

It should be clear who your supervisors are and how to reach them if there is an administrative emergency. If there is a medical emergency, who is your back-up?

As a new employee, you are entitled to a thorough orientation. This benefits not only you, but your patients and the facility. The hours you spend in orientation should be compensated at your usual rate.

Dress for Success

How you present yourself matters. Most of your professional and personal interactions will be brief. The less you talk, the smarter you'll appear.

As a new face, you will be judged in large part by your appearance. Some institutions, like the Mayo Clinic in Rochester, MN, enforce a strict dress code. Mayo clinic male physicians must wear ties and jackets except in the operating room. Most clinics and hospitals are more relaxed when it comes to workplace attire.

It's largely up to you how you dress. In my years of experience, a pressed long sleeve shirt, tie, and white coat still earn the highest level of respect from both colleagues and patients. Frankly, I hate wearing ties, but if a simple matter of style helps me do my job, it's worth the minor discomfort.

Scrubs are another alternative, but often reserved for surgical and procedural specialties. Depending upon the hospital, they may also be acceptable for ER physicians and hospitalists or nocturnists. A quick tour around the hospital during orientation should provide guidance on this question. If you are still uncertain about which apparel is allowed, ask whether the hospital has a written dress code.

Clothing choices for women are more complex, but any woman who has made it this far in the medical world knows what it means to "dress like a professional." A noncontroversial outfit is an easy way to facilitate acceptance into a new environment.

When I worked locum tenens in a busy neurology office outside Boston, I remember the day a new employee showed up in a surprisingly low-cut dress. While it was an attractive outfit, it was inappropriate for the setting. One of the other employees must have taken her aside, because she never wore anything like that again!

If you are a maverick and wish to challenge the system on your first day, do so at your own risk. Some people may appreciate a bold fashion statement, but it's more likely to generate gossip than compliments.

The Next Step

Tens of thousands of locum tenens physicians enjoy the freedom of scheduling their work and free time. The best opportunities come to those with availability, affability, good communication skills, flexible schedules (and personalities), inquisitiveness, ability to cooperate with recruiters and co-workers, and sound clinical skills. These qualities facilitate successful locum tenens work.

Where and when you want to work and for how long are up to you. Give these options some thought. Decide whether you want to combine vacation travel with a new assignment. Think about the work environment. Would you be more comfortable in a clinic or inpatient setting? Is the patient workload reasonable? Are you familiar with the EMR? How much on-call, if any, are you prepared to do? Are you keen to work with an "underserved" population?

Will you need a new state medical license? If so, think carefully whether you are likely to need that license again. If this is a one-time fling in Texas, it may not be worth the time and trouble to go through the licensing process (Chapter 9).

Lastly, how much compensation do you need to make an assignment worth your while? I turned down one long-term job because the daily rate just didn't cut it. Salary is negotiable, up to a point. When negotiations weren't fruitful, I chose another, better paying position.

In the final chapter that follows, you'll read first-hand accounts from experienced locum tenens physicians. By the time you finish, you'll know whether locum tenens is right for you.

Recommendations

New work environments create novel challenges that must be managed with confidence and skill. If you've done your research and arrive prepared, it should work out. Remember the "3 A's." In addition, I keep an index card in my shirt pocket with the words, "Be flexible."

CHAPTER 20

Tales from the Trenches

Locum Tenens Physicians Share Their Experiences

The vast majority of information in this book comes from my personal experiences and research. Of course, my own locum tenens adventures constitute only a small sample of situations that locum tenens physicians in a variety of specialties and in different clinical settings might encounter.

To expand the book's scope, I solicited in-depth comments from other locum tenens physicians. I asked them to share the circumstances that led them to choose locums, their best and worst assignments, as well as offer advice to physicians contemplating their first locum tenens assignment. During telephone interviews or emails, I communicated with specialists in anesthesia, emergency medicine, family medicine, infectious disease, neurology, orthopedic surgery, and physical medicine and rehabilitation. Physicians in all career stages contributed. (The sample does feature several neurologists, a reflection of my own physician network.)

A number of important themes emerged from these discussions. For example, locum tenens allowed physicians to pay off loans, save money to start a practice, return to clinical medicine after time away, escape a non-compete clause, earn a living while searching for the perfect job, travel across country, start a side business, expand their clinical experience to different patient pathology, regain control of the practice of medicine from hospital administrators, achieve a more satisfying work/life balance and semi-retire at an early age. Although this may represent sampling bias, not one physician expressed regret for trying locum tenens.

The fascinating tales that follow illustrate not only the positive aspects of locum tenens, but negatives as well. For example, freedom to travel was accompanied by loneliness, high salaries came without benefits, flexible assignments were cancelled at the last minute, and temporary assignments precluded long-term patient relationships.

The following comments were lightly edited for clarity and reprinted with permission.

A Newly Graduated Family Physician

I've been doing straight locums right out of residency. Initially, I wanted to set up my own practice, but I witnessed the horrors of being employed when I saw doctors who signed up for a hospital practice treated as patient mills, having to see a certain number of patients by the end of the day.

When a good colleague of mine mentioned locums, I thought I would give it a try. So far, I've worked in 4 states; an Indian reservation in Nevada, a VA in Louisiana, a homeless shelter in California, and a community health organization in Washington. The last assignment in Seattle was an awesome opportunity. They have a very diverse population, and I was able to work with people from Afghanistan, Nepal, and West and East Africa.

With locums, I was able to pay off my medical school loans. I have no complaints in terms of pay. I've also been able to develop my online Wellness Brand, www.thechefdoc.com. I went to culinary school and have a passion for the use of food in medicine. I'm using the power of the internet to teach people how to cook healthy meals, as well as self-improvement and motivation skills. I was able to build this brand on the side, and that wouldn't have been possible without the flexibility of working locums.

I haven't had any negative reaction or comments regarding my status as a locum tenens doctor. Someone might ask me why I don't go permanent, and I tell them from my point of view, I really value my work/life balance. It's important to take care of my own health and personal matters as well as my work.

I would say that locum tenens is probably the best option for newly graduated residents to narrow down what they want to practice and how they want to practice. I figured out what type of health care professional I need to be, but I wouldn't have been able to if I hadn't had the experience to work in different health care environments. Right now, I am looking at a perfect permanent position. I feel that I have more than enough experience to take on a permanent job.

A Family Medicine Physician Six Years Out of Residency

I left my first/only job after residency in November 2015. It was not a planned separation, so initially locums was my "placeholder" until I got another full-time job. I had done some "moonlighting" in residency (our residency actually encouraged it, which was very nice) and was familiar with locums work, especially since locums companies cover most of the ED in the state of Iowa and Missouri. I needed to be able to pay the bills and feed my family and finding full-time employment can take up to a year.

But it didn't quite work out that way. Life never really does. So, I am using locums now to build start-up money and pay down the horrid student loans. I live in Southeast Iowa, "30 minutes past the middle of nowhere." Really. Our closest Walmart and town with over 2,000 people is nearly 40 minutes away. I am currently working with two different staffing agencies, though one of them I like much better than the other.

I have negotiated my own locums work, but I do not like doing it, so I allow others to work on the hard stuff and I am okay with slightly less pay. My favorite part of locums is the variety and the fact that pay is negotiable.

I am very well trained in family medicine with emphasis in peds, psych, OB and trauma, so I can work just about anywhere. I am currently only doing ED work rather than urgent care or clinic work since I have a family and don't want to be away from them for super long stretches of time. Right now, I have three state licenses (Iowa, Missouri and Maine), and usually am not gone more than 10 days at a time, often shorter stretches if I can drive there. With the new interstate licensing options, I may very well increase the number of my state licenses.

My locums company has paid for my state licenses, DEA, state CSA, ATLS and others. They pay for my hotel, air travel, and mileage if I drive my truck or a rental car. They do all the credentialing for me so I don't have to. It's a very nice agreement and I enjoy it.

The big drawbacks for me are not having regular pay (comes in spurts with ED coverage), having to pay my own taxes, having to self-insure (we can't afford dental or vision insurance and are on ACA marketplace insurance) and being away from my family for several days. Part of this is a lifestyle choice and intentional.

I only make about half of what I did in my employed position, plus the loss of benefits. However, it has been HUGE for my mental health, and I am no longer feeling trapped by the bizarre world of medicine in this country. I still have HCAHPS and other nonsense, but because I am locums, it does not affect my pay in any way. I am able to provide very high level, quality care and my private practice runs on this principle, too. I am huge in patient education, and honestly the one thing ED nurses comment on is the amount of patient education I provide.

A Family Physician Eight Years Out of Residency

I'm an early/mid-career Associate Clinical Professor of Family Medicine at a California medical school. After casually speaking with many experienced doctors, some who were retired and some in their 40's and 50's, I noticed they all said the same thing-that they regretted not taking the opportunity when they were younger to see what was out there, learn more, and experience what it was like in different clinics treating different medical conditions before they settled down in their jobs and set roots down.

You know how we call interns, "R-1s"? I just consider myself an "R-12." I left my dream job that I was doing very well at just so I didn't miss the opportunity.

I have gotten to challenge myself at clinics where I was the only doctor five days a week. I have worked at clinics where they speak 86 different languages and have seen diseases that I had only read about in medical school. Seeing and learning to manage these has actually made me more confident and has made me a better teacher when I come back on my "off months" to teach at the university. Being able to help a clinic that really needs a doctor, while continuously learning and seeing new diagnoses in a different part of the country at the same time-it's a win-win situation.

I actually quit my full-time job, but fortunately I left on very good terms. The department has been very short-staffed so they are relieved and happy to have my help when I am around-it relieves a lot of strain both in precepting residents as well as shortening the wait for patients to get in.

The hardest thing for me has been seeing very difficult and complicated patients (i.e., drug-seeking patients, drug-dependent patients, depression/ anxiety) that you don't know and don't have enough time to develop a rapport with. It can also be disheartening to hear patients ask me if I'm "leaving in a few months, too."

I do plan to continue locums. Ideally, it would be three-month locums, three months home, three months locums, three months home. Would I recommend locums to someone in my situation? Absolutely yes.

A Mid-Career Rehabilitation Specialist

Out of residency, I worked as the medical director for a 1/2 billion dollar health-related internet portal and social network. Unfortunately, the company was before its time and ultimately did not succeed. After being out of the hospital setting for several years, I wanted to get my foot in again, and that's when the locum tenens idea came to me. I had done some volunteer work but was getting a little nervous that too many years had gone by and I would become unhireable. But I didn't want to commit to a full-time job.

In fact, a lot of places wouldn't hire me. They said "No, you've been out of it too long." It was kind of scary. Finally, I found a physician practice that hired me to work in the hospital. That worked out great. Once I was back in the system, I had no further trouble getting hired.

I do really well with locums, making about 30% more than I would in a permanent position. I think part of it is that there are no expenses, everything that you make is just free and clear. In addition, there's usually overtime pay because the facilities are understaffed. The salary varies wildly. If you negotiate well, you can really make a lot of money doing locums.

The uncertainty is the worst part. You have to be flexible. You have to know things can change at the last minute. It's very expensive for the facility to hire locum tenens physicians. It's really their last choice. A lot of times you have difficulty to schedule your life too far out, because if the hospital can find someone outside of a locums agency, they save 40% and will pull the assignment out from under you. It's true there are contractual protections against last minute cancellations, but in my experience, they are never enforced. The agencies don't want to make the hospitals angry. If the agency demands the cancellation fee from the hospital, the hospital will just go to another agency the next time.

These days, the treadmill in health care is set so high, you can't find a way to dial down. The only option you have is jump on or jump off. There's just no way to find a job that doesn't eventually entail 60-80 hours a week, and you're paid for 40. They want you to work on committees, go to meetings, and you want to, but you do all this uncompensated work, and in the end, you just burn out. You get compassion fatigue, and that's terrible. You never want to lose that for your patients, because you're just exhausted and frustrated, you're not the doctor you want to be.

It's also a great way to know other parts of the country. I thought New York City, where I trained, was the only place where I could see myself living. But then I spent time in Idaho, South Dakota, and Nebraska. It really expanded my horizons and made me realize that there are a lot of places you can be happy. Locums made me a better person, and a better doctor. I think that the millennials are going to love the locum tenens lifestyle once they discover it.

Locums is great because you can control your schedule. For me, locums is the only sane way to keep your compassion, and to stay actively involved in clinical work without being consumed by it.

A Mid-Career Neurologist

I've been doing locums exclusively for more than three years now (now and then for more than seven years), and have had long-term part-time clients where I've been covering one to three weeks at a time, with regular work every month. Maybe it's because neurology is in so short supply, but if you pick the right clients, it can work long-term.

Another Neurologist's View

I've been doing locums exclusively for the last two years, and fairly regularly for the 4-5 years before that. I'm not a great story teller, and nothing that interesting in my experiences so far, at least in my opinion. Have been lucky to work for places that have long-term need and accept long-term part-time work: Spokane, WA, at first, but over the last few years a couple of places in North Dakota-both places in ND had three neurologists recently, and still needed help at times, but one is down to a lone neurologist, and the other is down to no local person and only locums coverage due to the difficulty of recruiting and keeping docs there. They both seem like reasonable places, especially the one with no neurologists now.

I can do what I would consider a semi-retired work load 2-3 weeks a month (i.e., 13-14 patients per clinic day or round on 6-7 inpatients) and make close to a full-time salary for an overburdened full-time practice in California or Arizona (18-22 in clinic or rounding on 16+ inpatients daily).

Bridge to the Perfect Position

My permanent position soured after I had to testify in court against another physician. I had to leave and was having a hard time finding another job. I didn't want to make a mistake by making a hasty job decision just to have somewhere else to go. So I took a locums position for 6 months.

It worked. While working locums, I interviewed at several positions, and I found the perfect position, which I still have 16 years later. It's a physician-owned multispecialty practice, and we've been in the black for more than 100 years. The practice wanted me so much they even bought out my locums contract so I didn't have to complete the whole 6 months.

Although the locums assignment gave me the freedom to locate the perfect permanent job, it was not without problems. There was one doctor there who seemed to resent me. When the ER called me, I always went in. He never did. I always showed patients their x-rays on the white box, which happened to be outside his office. He never did. He told me I had to share in EEG reading, but then he wouldn't give me the password. I think I raised expectations of what constituted appropriate neurologic care, which irritated him.

I also felt very isolated. I left a big, beautiful house in Louisiana to live in some town I can't even remember the name of on the edge of Iowa. The locums agency put me in housing right next to the hospital. It was convenient, but very depressing. I remember that house wrens had built a nest inside the air conditioning unit. As soon as the little chicks were born, a crow attacked the next day and killed them all. It was really horrifying. It seemed to be like some kind of omen. Every Saturday, I bought myself fresh flowers to cheer the place up.

I would advise new locum doctors to make sure the living quarters are adequate and not some hell hole. If they have a pet, insist on bringing it with them.

On the plus side, the money was great! I had a bundle of money. The money was so fantastic that I ended up with $80,000 cash when I arrived at my new job after paying taxes.

I wish I had done locums sooner!

Locums Rescues an Infectious Disease Doc

I finished my training in infectious disease ten years ago in New York. Then I joined a multispecialty group in Ft Myers, FL. It was inpatient, outpatient, and busy, but I really wanted to be in Miami. Finally, I joined a very busy infectious disease group in Miami, but it was the wrong job for me. They worked me to the ground, and I was unable to reason with anybody. I was desperate to leave.

I called CompHealth in October 2015, just to ask a few questions, and they hired me on the spot for nine days of Christmas holiday coverage in South Dakota. Thanks to the cash I earned, I was able to take January off and start my own little practice in outpatient HIV in Miami. I only kept my office open a couple of mornings a week, because when I was home I just wanted to chill out and go to the beach. I was also able to work more locums. I went to Spartanburg, SC, for seven months, then a bunch of trips to Bismarck, ND.

One day last December, a representative of a nonprofit health care group appeared in my office. They wanted to take over my little Miami practice and give me a big, beautiful HIV practice. Their vision was in line with my vision, so I accepted. I am a little sad about leaving locums, but working with this nonprofit is a better long-term plan for me.

Locums was amazing. The only reason I stopped doing it is because the nonprofit made me a great offer. I'm very grateful for the locums experience, because I had to get out of that job in Miami.

I love traveling, meeting new people, and going to different places. The culture of medicine is so different region to region. It's really fascinating, and locum physicians are the ones that can appreciate that.

I'm not sure it's the best thing for someone just out of residency to do locums. I think it's better to join a practice and get some experience, because you could go somewhere and not know what to do. You kind of know what to do when you finish your fellowship, but not really. In infectious disease, you see some crazy things, and you need colleagues to talk with.

Locums work really makes you a very strong physician. I honestly feel like I could walk into any hospital anywhere, and work with anybody, and take care of any type of infection, except maybe some very complicated transplant patient. I could probably go into any smaller hospital and run the department, because that's what I did in North Dakota. It gives you the confidence that you can go to a strange place and just handle things. It's amazing.

A Mid-Career Anesthesiologist

I was in private practice for 11 years before starting to do locums. I did my job, working every third night and every other weekend on call, and didn't think anything of it. One day I realized I was working a lot, and for a diminishing paycheck.

I always believed, mistakenly so, that locums was somewhat of an "island for lost souls," if you had a drug problem, or other issues, it's where careers went to die. But then I needed to do locums because I was between jobs. To my surprise, I found out that locums is turning into the only safe haven left in the U.S. where physicians can practice the way they want to without having to deal with any of the administrative drama of practicing in the hospital.

There are just too many negative factors to being on hospital staff, especially the politics. When I was on staff, turnover time of the operating rooms was a huge issue. If any of the physicians dared complain, the techs or nurses would retaliate by "writing you up," saying that you were difficult to work with. Locums is a way of avoiding all that nonsense.

The first day at an assignment is always the most difficult, because you have to get used to a new system and new people, just so that everything can be smooth. If anyone doubts your abilities, it doesn't take long, one or two cases, and they are quickly convinced of your value.

The best thing about locums is that the honeymoon period never ends. Everyone is happy to see you, the patients, the staff, the administrators. After I started locums, I found that I was able to really just be myself, do the right thing for the patient, get paid a fair salary, and just actually start to enjoy why I became a physician in the first place.

A Financially Independent Anesthesiologist

It's tough to narrow down my "best assignment" to just one, since so much good has come from locum tenens jobs, including a couple full-time positions, good friends, Godchildren, and more. I'll go with one that gave us none of those things, but felt a lot like a three-week vacation. I had just finished residency, worked a one-week job, and taken the written anesthesia Board exam. My fiancée and I settled into a gorgeous condo in a resort that looked like an Italian villa in Naples, Florida. Every morning, I would walk to work at 0645, grab a bite of food on the way, and be back by the pool by a quarter after three. Each evening and weekend was free. I spent my per diem on sushi and other seafood. I earned as much in those three weeks as I did in the prior six months as a much-harder-working resident.

As to my "worst assignment," one summer I wanted to work in a particular part of a particular state. There were few options, but I was able to find a facility that could use my help as long as I worked at another facility 100 miles away for an equal amount of time. The pay was relatively low (and fixed) and the hours and responsibilities were far greater than had been described. At the time, I felt overworked and underpaid, but I was the one who put myself in a position for that to happen. I had little room for negotiation since I was asking to work instead of being sought out to fill a need. I did make some good connections with other physicians and the practice owner later made it up to me. It wasn't all bad, but at the time, I was not a happy camper. I was working so much that my fiancée went back home until our wedding because I was never around to see her.

In terms of advice to potential locum tenens doctors, I would offer the following: If you plan to do locums full-time and travel, get multiple state licenses, and get them all at the same time. Once you have numerous licenses, getting each additional license is costly and a bureaucratic nightmare. This may become less true with the new interstate compact, but most states are not yet participating. Also, remember that you can work in any VA or IHS hospital nationwide with one state license.

Look for the position you want, and work with the agency that offers it. (I took the opposite approach, getting credentialed with several popular agencies, and never worked with any of them.)

Learn where to find locums jobs within your specialty. The best source may be a specialty specific job site, a state society, or through word of mouth. If you're only looking locally, be sure you're not in violation of a non-compete agreement.

Get everything in writing.

If you don't ask, you don't get. For example, I've gotten a rental car even though I drove to the site (my wife used our car on site). I've gotten a second adjacent hotel room for our infant son and dog. In a longer-term job, we were granted a fully furnished 3-bedroom apartment in a dog-friendly downtown building. Without special requests, we wouldn't have gotten any of these benefits.

Do your job well and approach everyone you meet in a friendly, respectful manner. I've been asked to return or stay on full time at nearly every job I've worked. You never know when you might want to take someone up on such an offer.

A Neurologist Waits Out a Non-compete Clause

I am in the last decade of my career. I only started doing locums one year ago. I graduated from my neurology residency in 1985 and then went back and did a fellowship from 88-89. I would like to work for about 8-10 years more. So, since I've been practicing a little over 30 years, I guess you could say I'm in the last ¼ of my career. I'm not planning to retire in the very near term.

I work in the desert Southwest and also in the Midwest. I live in the Pacific Northwest. I would consider an assignment in any of the 50 states including Hawaii and Alaska.

Like everyone else, I was getting emails and even phone calls from locums agencies several times a week for some time. I began to consider locums when I was looking for an alternative to a job that was economically very poor. I think I met my locums agency representative through LinkedIn. We talked for over a year before I got to the point of resigning from my permanent job and starting locums employment.

I was very skeptical at first, but eventually the combination of my worsening financial position in the job I left (80% overhead/80% Medicare and Medicaid payer mix, and 17 days of long-term EEG call a month) made me decide to make a change. I have a 2-year non-compete clause so it was either move or do locums. My locums rep gained my trust. I did a pre-credentialing application and so it was pretty easy to make the transition.

The feeling of regaining control of my professional life and to be compensated fairly are the best things about locums for me. I decided to establish a single member LLC and pay myself and my benefits through the LLC. I have a bookkeeper, an accountant, and an attorney. I am able to deduct the full costs of my practice including CME, IRS allowed per diem, health insurance, and retirement contributions as business expenses. Because of the IRS treatment of retirement in an LLC, I am able to contribute dramatically more than I would be able to do as an employee.

The worst thing about locums is the feeling of not being in control of my work schedule. I have had recent experiences where two rather substantial recurring assignments were cancelled at the last minute before I started work. In both, I had a signed contract and one was within 14 days of starting. In the first, the hospital hadn't processed my credentialing application and has refused to pay a cancellation fee. In the second, I also have a signed contract and the cancellation fee for the first assignment has been promised, but I wonder if I'll ever see it ($13K).

I also have found that I get sick when I fly so much. I have contracted strep that turned into nasty pansinusitis and a really severe adenovirus infection related to air travel after tending to a medical emergency on a flight in a passenger that I thought was going into septic shock.

Moreover, I have become increasingly concerned about small carbon particle and organophosphate (cresyl phosphate) exposure in airliners (see www.aerotoxic.org). I now fly about 4-8 flights a month and generally feel really bad the day after I fly even when I don't get sick. I'm now trying a Cambridge mask, an N-99 personal HEPA filter with a carbon fiber inner layer (for VOC's), to see if it helps. In a limited experience so far, I would say yes, it does. I wipe down my seat and the tray table with antimicrobial wipes, and I bring my own water and coffee on the plane.

I don't like being away from home. I used to get anticipatory homesickness a couple of days before flying out and feel pretty homesick for a day or two after I arrived in the assignment. That has gone away although I admit it's a pretty lonely life at times.

Financial aspects totally depend on how much you work. On a surface analysis, it would seem to be MUCH better compensated. The LLC grosses about as much in 12 days as I made in a month in my previous job's guarantee. But it's an apples and oranges comparison. Self-employment/payroll tax, health care premiums, and larger retirement add about $4,500 per month to the debit side of the ledger. My costs on the road also seem higher than at home.

I had been working 19 days a month until September. That was to have continued from October to the end of the year (I took a vacation for two weeks in Sept). However, my Oct-Dec assignment for seven days a month was cancelled while I was on vacation. Now I have a lot of time off, but also the LLC will show a monthly loss through the end of the year.

The absolute chaos in the health insurance market is a major issue going forward. I may decide to take an employed job in 2018 to get health insurance.

Nonetheless, I absolutely would recommend locums for someone in my situation. I think that locums is a great option if you're stuck in a bad job and don't want to move. It removed the feeling of being trapped in a bad job and has restored some sense of financial security.

The effect on my career has been positive overall. I've made good connections at two academic medical centers. I'm probably going to take on editing a textbook on integrative neurology. I've been able to regain my perspective about where I want to go with the last 25% of my career. Those are all precious gifts.

I am not sure how much longer I will continue locums, but the only regret I have is not having done it sooner in my career.

A Family Practice Doctor Works Locum Pre-Retirement

I graduated medical school in 1984, did a residency in family practice and have been in active practice since 1987. The hospital bought my practice in 1995 and I stayed until 2012, when I quit.

The MBAs of the world have destroyed family medicine. There's no one going into it for good reason. I believe in direct patient care. The payment model is unsustainable for family medicine. The Academy of Family Physicians is pretty much impotent. It has left a bad taste in my mouth. I think it needs to be unionized.

In 2012, I decided to take off and go to New Zealand for seven weeks. When I got back, I started working with CompHealth. I did locums 30 miles away for three months, then Presque Island Maine, in the winter, then two locums stints in Hawaii for Kaiser. I learned to scuba dive and took three months off between jobs.

Locums has a beginning and an end. I work hard. I try and do what is needed, try not to say no. If you are done after four months, even if it wasn't ideal, it's done. You can be happy even in a system that's difficult, as it ends. The best things about locums is I get to do medicine.

The culture and vibe in Hawaii was pretty cool. You get to benefit from being in a different geographic area when it comes to interesting cases, too. Some of the stuff I saw was sports injuries, ear infections, and water injuries.

But even in Hawaii, you get tired of it. It doesn't matter if you are in Maui if you are in the office all day. It is an island, and you are looking forward to getting off. My practice was 25 miles from the nearest hospital, so it was challenging. My wife is a physician, but she wasn't with me because she couldn't leave her job, so there's some loneliness, but you keep yourself busy.

I work urgent care now, staying near my home, about 18 hours a week. I'm not going to get another regular job.

A 73-Year-Old Neurologist

I've worked locums at various hospitals with various models. I think it all depends on the size of the hospital. For a large (>400 bed) hospital with a busy ER, 24/7 is more than I can do anymore at my age. I need to get more uninterrupted sleep.

Teleneurology has made a big difference at some hospitals I've worked for. There, you do a 10-hour shift of 7 am to 5 pm, and nights and weekends are covered by teleneurology and a robot. At smaller hospitals, 24/7 is more viable as you won't be getting several calls during the night. At least in the locum market, it is hard to get neurologists to work a large hospital 24/7. Recently, a 450 bed hospital I worked for has had to get teleneurology for 10 pm to 6 am or else go uncovered as no neurologist was willing to work 24/7.

A Semi-Retired Orthopedic Surgeon

I learned about locum tenens from an email when I was full-time faculty about 20 years ago. I've been working locums more than 10 years, and work all over the country. I have five active medical licenses from states in the north, south, east and west. When the assignment ends and staff and other health care providers have appreciated the quality of your work, you know that you have made a difference.

Locum tenens allows control over a professional and personal schedule, and it allows the freedom to plan around other professional and personal commitments. As Community Faculty affiliated with the local medical school and residency program, I have opportunities to present didactic lectures to the orthopaedic residents. I also spend time volunteering at an indigent clinic. It can equal salary in an academic or hospital group setting if you are willing to make a commitment on average of 2-3 weeks a month.

Locum tenens has broadened my breadth and depth of experience with patients, pathology, cultures and communities. It has allowed learning and education in areas of medicine and orthopaedic surgery that otherwise may not have been available. It has allowed new professional relationships with other wonderful locum tenens providers and other excellent local physicians from other specialties. It has enhanced my overall professional life experience!

Frank Talk On a Long and Varied Locums Career

My first locums job was in 1981, when I was going to two conferences, one in New Orleans, one in Las Vegas. Back then, conferences were Monday-Thursday, so I decided to do something in-between, rather than fly home and fly out. I arranged an Illinois license (much easier back then), bought an emergency kit (from Banyan), with meds, intubating equipment, etc., and showed up on Friday. As I was driving down from O'Hare, I noticed a sign that said, "Trauma Center 10 miles," and thought, "Great, at least it will be a busy place."

When I got there, I found out that I was the trauma center for this 4 bed ED in Quincy, IL, a town with a prison, and doctors who signed out to the ED for the weekend. I worked from Friday afternoon through to Monday morning for $35 an hour, seeing all comers, and covering the house. I intubated a 99-year-old in the ICU, handled a multiple trauma (shipped out), child abuse, rape case, and a few other high intensity issues.

I also got my first exposure to ignorant antisemitism as, during the down time when the nurses found I was Jewish, asked me when my horns and tail were removed. It was a Catholic hospital, in a rural part of Illinois. I ended up going back a couple of times, basically because the money was fairly good, for its day, and the workload was about a patient an hour.

Several years later, I was looking for a part-time weekend locums assignment (I was doing occupational medicine at the time, Monday-Friday), and got a job two hours away. I went up on a Friday and came back on a Sunday. After my first shift, they realized I wasn't just someone who could handle a fast track, and they offered to buy out my locum company's contract. I worked there for 4 years and returned 10 years later for about 6 months.

I worked for a national staffing company and took an assignment in a rural part of Pennsylvania, fully resolving never to go back after the first time, since they put me up in an abysmal motel, where I didn't feel safe. The job, and the people were fine, however, and I was persuaded to go back, with better accommodations. Ultimately, I found out that the hospital was establishing an ER residency, and I was recruited for that, and ended up working there for 6+ years.

I was screwed by the locums company, however, when a nuisance suit that occurred on my first day of working locums was settled without my consent, after I expressly demanded that it not be settled. The suit affected my career very little, except to the extent that I didn't trust locum tenens organizations and my then current employer. I was thrown under the bus by them, as were two of the attendings involved in the medical care of the patient.

Most other locums jobs I've had have been routine - we'll pay you this amount, for that much work, and no problems.

A Retired ER Doc

I loved locums. I have four kids and needed more flexibility. The people were so delightful and it gave me a break from the usual.

We were practically required to moonlight in rural EDs and at the other trauma center in our city during residency. My first locums assignment was in a small rural ED in the Deep South. I arrived for my shift at 5:45 am one Sunday, and this sweet old nurse Lucille with soft white hair and gentle drawl said, "Doctor, we have your room all ready for you. Just go back to bed and sleep until church is over-no one comes in until after lunch on Sundays."

Oh, and she added, "Give me your breakfast order and I'll be sure we have it ready for you." And they did. I loved that place. It is in danger of closing now.

Although I worked for a long time after residency at a local hospital, I did some locums a few years down the road in some rural areas. They were so wonderful to me. When I quite practicing, I once Googled myself and saw that my only review on Yelp or whatever was someone who said, "I heard Dr. X left practice. I wish she would come back. She was so wonderful to the people here." I felt good about that.

On the other hand, there were some less idyllic times. I used to work a shift from Friday at 7 pm to Monday at 7 am and once had to call my husband to pick me up halfway home because I was too tired to drive. You know how that goes. At that locums job, I had to round on all the inpatients every day for the local doc. So that was three times during the weekend, sometimes 40 patients. I also missed an acute abdomen once. I still think about it, although I consulted surgery eventually and nothing catastrophic happened. Still, I was tired that day and had a bad attitude.

Once, there was a time I had hired a new gardener. That weekend, I was staffing the ED locally and the nurses came to me and said, with that telling eye roll, "The people in room #9 are drug seekers, in here all the time." I walked in and there was my new gardener. He looked extremely shaken up and said, "Oh, Ms. X, I didn't know you would be here."

My ex still hears from some of the people I saw in rural EDs. Seriously. I moved, but they call him and check on me. Sweet. I love people. That's why I became a doctor, sharing that crazy love.

Recommendations

Take these tales to heart. They are all true. Now it's time to decide whether you would like to add your own story!

REFERENCES

Chapter 1

Barron L. *Locums Lifestyle: Take Control of Your Life and Make Great Money.* Locumslifestyle.org, 2013.

Physicians Foundation. 2016 Survey of America's Physicians. Physiciansfoundation.org.

Staffcare. 2017 survey of temporary physician staffing trends. AMN Healthcare.

Wilner A. Locum-Tenens-The job you already have. Medscape.com, October 24, 2016.

Chapter 2

Locumstory.com (CHG Healthcare).

Staffcare. 2017 survey of temporary physician staffing trends. AMN Healthcare.

Chapter 3

Wilner A. *Bullets and Brains.* CreateSpace Independent Publishing Platform, North Charleston, South Carolina, 2013.

Chapter 4 (none)

Chapter 5

Lenhart A. Doctors aren't following their own recommendations for parental leave. Slate.com, February 19, 2018.

Mechaber HF, Levine RB, Manwell LB et al. Part-time physicians...prevalent, connected, and satisfied. *J Gen Intern Med* 2008;23(3):300-3.

Pinola M. How to get rid of 'mommy guilt' (or daddy guilt) once and for all. Lifehacker.com, January 15, 2015.

Temple J. Resident duty hours around the globe: where are we now? *BMC Medical Education* 2014;14(Suppl 1):S8.

Terry K. For first time, under half of physicians own their practices. Medscape.com, June 2, 2017.

Young A, Chaudhry HJ, Pei X et al. A census of actively licensed physicians in the United States, 2014. *Journal of Medical Regulation* 2015;101(2):8-23.

Welch A. Number of U.S. women taking maternity leave unchanged for two decades. CBS News, January 19, 2017.

Chapter 6

Bernat JL. How can neurologists avoid burnout? *Neurology* 2017;88:1-2.

Brann J. Stemming physician turnover. *Nevada Daily Mail*, August 3, 2017.

Lake R. How much income puts you in the top 1%, 5%, 10%? Investopedia.com, September 15, 2016.

Peckham C. Medscape national physician burnout and depression report 2018. Medscape.com, January 17, 2018.

Roitman L. Barely scraping by with a $250,000 salary in Silicon Valley. Hackernoon.com, April 27, 2017.

Staffcare. 2017 survey of temporary physician staffing trends. AMN Healthcare.

Valcour M. When burnout is a sign you should leave your job. *Harvard Business Review*, January 25, 2018.

Wilner A. Combating burnout: A neurologist's perspective. Medscape.com, February 23, 2017.

White Coat Investor. Moving the goalposts-attack of the lifestyle creep. Whitecoatinvestor.com, July 27, 2016.

Chapter 7

Burling S. More doctors are practicing past age 70. Is that safe for patients? Philly.com/health, September 8, 2017.

ReachMD.com. Practicing medicine after retirement. https://reachmd.com/programs/your-career-in-healthcare/practicing-medicine-after-retirement/6978/.

Saley C. Survey report-Physician views on retirement. CompHealth, July 31, 2017.

Silver MP, Hamilton AD, Biswas A, Warrick NI. A systematic review of physician retirement planning. *Human Resources for Health* 2016;14:67.

Staffcare. 2017 survey of temporary physician staffing trends. AMN Healthcare.

Chapter 8

Wilner A. Locum tenens-The job you already have. Medscape.com, October 24, 2016.

Chapter 9

Young A, Chaudhry HJ, Pei X et al. A census of actively licensed physicians in the United States, 2014. *Journal of Medical Regulation* 2015;101(2):8-23.

Chapter 10 (none)

Chapter 11

Cash CD. Risk management issues when taking locum tenens assignments. *Innovations in Clinical Neuroscience* 2017;14:60-62.

Kane CK. Policy Research Perspectives. Medical liability claim frequency: A 2007-2008 snapshot of physicians. American Medical Association, 2010.

Kreimer S. Six ways physicians can prevent patient injury and avoid lawsuits. *Medical Economics*, December 10, 2013.

Sklar DP. Changing the medical malpractice system to align with what we know about patient safety and quality improvement. *Academic Medicine* 2017;92:891-4.

Studdert DM, Mello MM, Gawande AA et al. Claims, errors, and compensation payments in medical malpractice litigation. *NEJM* 2006;354: 2024-2033.

Wilner AN. An Epic adventure. Medscape.com, January 20, 2013.

Chapter 12

Weger D. Going bare-are doctors required to have malpractice insurance? Gallaghermalpractice.com.

Chapter 13 (none)

Chapter 14 (none)

Chapter 15

Muchmore M, Duffy J. The best cloud storage and file-sharing services of 2018. *PC Magazine*, January 23, 2018.

Poole O. "Curse. It's the luck of superman." *The Telegraph*, December 9, 2002.

Chapter 16

Medical Economics. The top 100 EHR companies. October 25, 2013.

Sinsky C, Colligan L, Li L et al. Allocation of physician time in ambulatory practice: A time and motion study in 4 specialties. *Ann Int Med* 2016;165:753-760.

Wilner A. An Epic Adventure. Medscape.com, January 20, 2013.

Wilner A. Radiology rounds-Where have they gone? Medscape.com, July 5, 2010.

Chapter 17 (none)

Chapter 18

Staffcare. 2017 survey of temporary physician staffing trends based on 2016 data. AMN Healthcare.

Chapter 19

Cavanaugh C. 5 Qualities a recruiter looks for in a locum tenens candidate. Barton Associates, October 5, 2017.

Murray SD. The 3 A's of medicine. *MD Magazine,* Intellisphere, LLC, 2017.

Saylor E. 7 Skills hospitals are looking for in a physician. https://comphealth. com/resources/7-skills-hospitals-physician/, June 24, 2016.

Wilner A. An Epic Adventure. Medscape.com, January 20, 2013.

Chapter 20 (none)

APPENDIX 1

Glossary of Acronyms

AAMC-Association of American Medical Colleges

ACA-Affordable Healthcare Act

ACCME-Accreditation Council for Continuing Medical Education

ACGME-Accreditation Council for Graduate Medical Education

ACLS-Advanced Cardiac Life Support

APP-Application

ASAP-As Soon as Possible

ATLS-Advanced Trauma Life Support

BLS-Basic Life Support

CAQH-Council for Affordable Quality Healthcare

CAMH-Comprehensive Accreditation Manual for Hospitals

CME-Continuing Medical Education

COM-LEX-USA Comprehensive Osteopathic Medical Licensing
                    Examination

CPA-Certified Public Accountant

CPR-Cardiopulmonary Resuscitation

CPCS-Certified Provider Credentialing Specialist

CPMSM-Certified Professional in Medical Services Management

CSA-Controlled Substance Act

DOD-Department of Defense

DSLR-Digital Single-Lens Reflex Camera

ECFMG-Educational Commission for Foreign Medical Graduates

ED-Emergency Department

EM-Emergency Medicine

EEG-Electroencephalography

EMG-Electromyography

EMR-Electronic Medical Record

EMTALA-Emergency Medical Treatment and Labor Act

EPIC-Electronic Portfolio of International Credentials (not to be confused
    with the widely used electronic medical record "EPIC.")

ER-Emergency Room

FBI-Federal Bureau of Investigation

FCVS-Federation Credentials Verification Service

FLEX-Federation Licensing Examination

FMG-Foreign Medical Graduate

FSMB-Federation of State Medical Boards

GSA- U.S. General Services Administration

HEPA-High Efficiency Particulate Absolute

HCAHPS-Hospital Consumer Assessment of Healthcare Providers Systems

HIPAA-Health Insurance Portability and Accountability Act

HIV-Human Immunodeficiency Virus

HSA-Health Savings Accounts

ICU-Intensive Care Unit

IHS-Indian Health Service

IMLC-Interstate Medical Licensure Compact

JCAHO-Joint Commission on the Accreditation of Healthcare
        Organizations

LLC-Limited Liability Company

LT-Locum Tenens

MBA-Masters of Business Administration

Mbps-Megabits Per Second

MLS-Multiple Listing Service

MSP-Medical Services Professional

NALTO-National Association of Locum Tenens Organizations

NBME-National Board of Medical Examiners

NCQA-National Committee for Quality Assurance

NPDB-National Practitioner Data Bank

OCR-Office of Civil Rights

PC-Personal Computer

PDF-Portable Document Format

PHI-Protected Health Information

PMP-Prescription Monitoring Program
PMRS-Prescription Monitoring and Reporting System
U.S.-United States
USMLE-United States Medical Licensing Examination
VA-Veterans Administration
VOC-Volatile Organic Compounds

APPENDIX 2

Locum Tenens Agencies

- AB Staffing Solutions
- Advantage Locums, LLC
- Alliance Recruiting Resources
- AIMS Locum Tenens
- All Medical Personnel
- All-Star Recruiting
- Alliance Recruiting Resources, Inc
- Alumni Healthcare Staffing
- AMS Locums/Assurgent
- Apex HealthSync
- Aya Locums
- Barton Associates
- BlueForce Healthcare Staffing
- Cancer Carepoint
- Catalina Medical Recruiters
- Catapult Healthcare
- CompHealth
- Concorde Staff Source
- Consilium Staffing
- CoreMedical Group
- D&Y Locums
- Delta Locum Tenens
- Docs Who Care
- DocsDox.com
- DOCS4ADAY LLC

- DR Wanted.com LLC
- Elevate Healthcare Consultants
- Eskridge & Associates
- FCS, Inc.
- Floyd Lee Locums
- Fusion Healthcare Staffing
- Global Medical Staffing
- Goldfish Locum Tenens
- Harris Medical Associates, LLC
- Hayes Locums
- Healthcare Connections, LLC
- Health Mount LLC
- HealthTrust Locums
- Honor Medical Staffing
- IMP Locum Tenens, LLC
- Integrated Resources
- Integrity Locums, LLC
- Interim Physicians
- Jackson & Coker
- KPG Healthcare
- KPS Locums
- LocumTenens.com
- Locum Connections
- Locum Jobs Online
- Locum Leaders
- Locum Life LTD
- Locum Physicians, LLC
- Locums Unlimited, LLC
- Lotus Medical Staffing, LLC
- Lucidity
- Maxim Physician Resources
- MD Staff Pointe
- MDspots.com
- MedCare Staffing
- Med-Link Staffing,Inc
- Medestar

- Medical Doctor Associates
- Medical Search International
- Medicus Healthcare Solutions, LLC
- Medifield Staffing, Inc.
- MedPartners Locum Tenens, Inc
- Medstaff National Medical Staffing
- Mint Physician Staffing
- Monroe and Weisbrod
- National Medical Resources, Inc.
- Nortek Medical Staffing, Inc.
- Next Medical Staffing
- Nomad
- Onyx M.D.
- Pacific Companies
- PhysiciansPRN
- Physician Resources, Inc.
- Pinnacle Locum Tenens, LLC
- Preferred Health Care
- Premier Physician Services, LLC
- ProCenture Healthcare Solutions
- Provide Locum Tenens
- Quest Healthcare Solutions
- RomanSearch, Inc.
- Rx Relief
- Sage Staffing
- SBG Healthcare
- Smart Physician Recruiting
- Staffcare
- SUMO Medical Staffing
- Synergy Physicians, Corp.
- T-PSYCHIATRY
- The Execu-Search Group
- The Locums Company, LLC
- Ultimate Locum Tenens
- VeloSource
- Vista Staffing Solutions

- Vitruvian Medical
- Wapiti Medical Staffing
- Weatherby Healthcare
- Western Healthcare

AFTERWORD

I hope you enjoyed reading this book as much as I enjoyed writing it. I've tried to supply essential information for locum tenens physicians sprinkled with a dash of entertainment.

Our current health care nonsystem presents formidable obstacles to physicians who aspire to fulfill their calling. Administrative changes that benefit locum tenens physicians, such as the interstate medical licensure compact, have been slow to evolve. There is still much to be done to minimize the medical/legal bureaucracy that limits physician availability and undermines patient-doctor relationships.

As our population ages and requires additional medical care, the need for locum tenens physicians will continue to increase. People will always need capable and compassionate doctors.

While locum tenens may not be the ideal way to practice medicine, perfect solutions to modern problems are tough to find. For tens of thousands of physicians, locum tenens offers a pragmatic path to practice medicine on their own terms.

If you found this book helpful, or even if you didn't, please share your thoughts in an Amazon.com review. To read my blog and receive updates on future books, visit my website: andrewwilner.com.

Thank you for reading and the good work you do.

Andrew Wilner, MD
Memphis, TN

INDEX

Made in the USA
Monee, IL
27 October 2020